D1014744

Why Air Forces Fail

Why Air Forces Fail

THE ANATOMY OF DEFEAT

Edited by Robin Higham and Stephen J. Harris

THE UNIVERSITY PRESS OF KENTUCKY

Publication of this volume was made possible in part by
a grant from the National Endowment for the Humanities.

Copyright © 2006 by The University Press of Kentucky

Scholarly publisher for the Commonwealth,
serving Bellarmine University, Berea College, Centre
College of Kentucky, Eastern Kentucky University,
The Filson Historical Society, Georgetown College,
Kentucky Historical Society, Kentucky State University,
Morehead State University, Murray State University,
Northern Kentucky University, Transylvania University,
University of Kentucky, University of Louisville,
and Western Kentucky University.
All rights reserved.

Editorial and Sales Offices: The University Press of Kentucky
663 South Limestone Street, Lexington, Kentucky 40508-4008
www.kentuckypress.com

10 09 08   5 4 3 2

All photographs, unless otherwise noted, are from the authors' collections.

Library of Congress Cataloging-in-Publication Data

Why air forces fail : the anatomy of defeat / edited by Robin Higham and
Stephen J. Harris.
p. cm.
Includes bibliographical references and index.
ISBN-13: 978-0-8131-2374-5 (hardcover : alk. paper)
ISBN-10: 0-8131-2374-7 (hardcover : alk. paper)
1. Air forces—History—20th century. 2. Aeronautics, Military—
History—20th century. 3. Military history, Modern—20th century.
I. Higham, Robin D. S. II. Harris, Stephen John.
UG625.W59 2006
359.4'8—dc22   2005030640

This book is printed on acid-free recycled paper meeting
the requirements of the American National Standard
for Permanence in Paper for Printed Library Materials.

Manufactured in the United States of America.

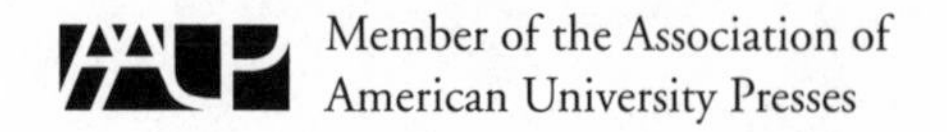

Contents

Acknowledgments

The editors wish to thank Marolyn Caldwell for her role in communicating with the authors and editors and in producing the final manuscript and Dr. Michael Lambert for the computer graphics.

Introduction

Robin Higham

Rather than being an exhaustive effort to examine the fall or defeat of every air force, this is a limited study in which we asked experts in the field to examine archetypal examples from which worthwhile conclusions could be drawn. This means, of course, that we had some ideas about what contributed to such failures before any of the authors put pen to paper. Admittedly, the notions of "defeat" and "fall" are applied very loosely here, and some might suggest that it would have been more useful to address the reasons why air forces failed to accomplish what their national command authorities asked them to do. Colleagues who were not involved in the project proffered many of their own generic causes for defeat (and victory) in air battles, campaigns, and wars. The ease with which such a priori conclusions were drawn led us to doubt the validity of our project more than once: if things were that clear, what could this book add to anyone's understanding? For example, given the importance of technology, it is intuitive that, other things being equal, the technically more advanced—more modern—air force should beat the less advanced opponent almost every time, provided its personnel can sustain the effort. But it is the nuances of those qualifiers—"other things being equal," "should," "almost every time," "provided"—that our authors address.

Defeats of air forces are both comparable and contrastable, and various tools can be used for analysis. A good place to start is Alfred Thayer Mahan's six criteria for success: the borders of a country, its physical conformation, its resources (in this case, the aircraft industry and its ancillaries), the size of the population, its characteristics, and the nature of the government. For air forces, additional criteria are relevant: the location and sufficiency of the bases from which they operate, the terrain over which they fly, the capabilities or efficiency ratings of their machines and weapons, and how well policy makers understand their assets, timing, and limitations—that is, the management factor in war. Put another way, air forces reflect their geographic and historical backgrounds, including the national character and cultural perceptions, the industrial and technological base, the political and diplomatic milieu, and logistical support. Their success or failure depends on contemporary strategy, operational art and tactics, and training.

Of the major twentieth-century airpowers, Britain, Japan, and the United States enjoyed isolation by the sea, while France, Germany, Italy, and Russia all had at least one long land frontier—although rugged territory on the border can make the effective use of airpower difficult. The size of countries also affords some protection. The most notable example of the effect of the frontier's conformation on airpower occurred in 1940, when German Panzers overran airfields in France and caused the downfall of the French armed forces. In contrast, the English Channel saved the Royal Air Force (RAF) in the summer of the same year. In the Soviet Union in 1941, space was a defensive asset. Natural resources too have a profound effect on an air force's ability to survive. These include not only oil, wood, and aluminum but also man- and woman-power.

For air forces, the size of the country's population is not as vital as its characteristics. Nor is the ability to recruit and train pilots and other aircrew the only factor. Because of the highly technical nature of the air service, the ability to obtain not just ground crews but also people to work in manufacturing is critical. As the Soviets showed in the Great Patriotic War (1941–1945), efficient use of people power required that designs be adjusted to the materials available and to the skills and know-how present in society.

In France, by comparison, technological advances outran the country's ability to foresee the need to recruit and train the necessary aircrews and mechanics in the 1930s. Thus, when the French air force went to war in 1939, it could not use all the modern machines its aircraft industry was producing. Similarly, RAF Bomber Command had to count many of its aircraft as unserviceable in 1944 because of a lack of radar mechanics—in a society where technical education was woefully inadequate.

The character of the population is also directly related to the way the government functions. Here, there are two crossovers: (1) the nature of the economic and business systems and their managerial mentality, and (2) whether these organizations have come out of a depression or a boom. Because government is generally composed of people from the same social establishments, their sense of planning, priorities, and procurement, as well as the creation of grand strategies and defense policies, are affected in similar ways. In Britain, for instance, in the interwar years (1919–1939), members of the government came from either the ruling class, including business leaders, or from labor, led by deserters from the capitalist elite. All were affected by the aftermath of the Great War and were involved in disarmament, so that for political reasons, the air arm was absorbed in advocating a non–trench warfare defense while devoting much of its effort to an aerial continuation of nineteenth-century frontier warfare. In contrast, in the aftermath of the Russian revolutions and the civil war, the proletariat leadership of the Soviet Union reassessed the position of the motherland, which they saw as surrounded by imperialist enemies. Thus, their first concern was the safety of the state. This led to the remilitarization and industrialization of the Soviet Union, planned in the period 1924–1928 and then inaugurated with Stalin's infamous (but managerially correct) Five-Year Plans. Parallel to these developments was the reestablishment of the aircraft industry and of the Red Air Force to such an extent that it emerged by 1934 as the world's largest and most potent air arm. And when war came, it rapidly rationalized its designs and production to produce what was needed for overwhelming air superiority and support of the Red Army.

The severest test of government is whether, in times of war, it can integrate a viable grand strategy with available resources, manpower, and the nature and vulnerability of both the enemy and its own vital resources, including lines of communication. Classic failures in this respect were those of France in the interwar years, Stalin in 1941, the United States in the Pacific in 1941–1942, and Göring's disdain for the RAF, which led to his failure to prepare the air defense of the Third Reich.

The air arm is particularly vulnerable because of the relatively small size of its fighting edge and because of the fragility and targetability of its bases, whether airfields or aircraft carriers, as well as its complex, highly technical, and inflammable logistics pipeline and salvage and repair structure. Moreover, the security of these fixed installations often depends on other services, so that in the end, air forces may fall because either the army or the navy (or both) failed to provide the secure base called for in everyone's principles of war.

A multitude of questions were suggested in the original information sent to the contributors to stimulate their thoughts and encourage them to go below the technical, tactical, and political surface. These included the influence of prophets, parsimonious political pacifism, preconceptions of all sorts and at all levels, personalities, purges, racism, doctrine, understanding of operational art, wastage and consumption, wartime dilution at all levels, preparations for war (including economic and fiscal plans), lessons of the last war, demobilization, and the realism of exercises and war games.

As the essays show, the failures of air forces have had multiple, in-depth causes, some in common and others unique, depending not so much on the date as on the circumstances, not so much on the technology as on the prescience, understanding, and management of that resource, and on the recognition of the importance of airfields and the support services—from mechanics to supply sergeants. In all cases, the conception of war, enemy capabilities and actual actions, and response to realities played their part.

Defeated air forces fall into three categories: (1) those that never had a chance and whose defeat was part of a national shame (the "dead ducks"), (2) those that had initial success but eventually failed (the "hares" that ultimately lost the race), and (3) those that suffered initial disasters but were victorious in the end (the "phoenixes"). In the first category are the German air force in World War I (chapter 4 by John H. Morrow Jr.), the Russian air force in 1914–1941 (chapter 9 by David R. Jones), the Polish in 1939 (chapter 1 by Michael Alfred Peszke), the French in 1933–1940 (chapter 2 by Anthony Christopher Cain), and the Italians in 1940–1943 (chapter 5 by Brian R. Sullivan). The Luftwaffe (chapter 7 by James S. Corum) and the Japanese (chapter 6 by Osamu Tagaya) represent the hares; they might have had a chance, but both suffered defeat as a result of grand strategic miscalculations. The phoenixes are the Argentines (chapter 8 by René De La Pedraja), the Arab air forces (chapter 3 by Robin Higham), the Royal Air Force in Norway, France, Greece, and Malaya in 1940–1942 (chapter 11 by Robin Higham and Stephen J. Harris), and the United States in the Pacific in 1941–1942 (chapter 10 by Mark Parillo).

Though the defeat of air forces may be rapid, leading to the sense that they are fragile, the simplicity of the ultimate end is only the tip of an iceberg composed of many complex, interrelated causes, as the chapters that follow illustrate. In the end, two questions have to be asked in each case: Did loss of air superiority, if it ever existed, cause the collapse of the nation's defenses? And was that the sole cause?

How air forces fall sometimes depends on whether the defeat occurs at the beginning or end of the conflict. Moreover, as will be shown, the defeat of air forces rarely has a single reason. It might be caused immediately by a shortage of aircraft and inadequate command vision. It may be due to a lack of personnel, both air and ground crews, or to a shortage of spare parts, tools, or fuel. In addition, a paucity of motor vehicles, immobile air stores parks, vulnerable fuel storage and distribution facilities, and inadequate repair, salvage, rebuilding, and maintenance organizations can contribute.

A principal immediate factor, of course, is loss of air control to the enemy, as well as the lack of a warning system. Loss of air superiority, as the Germans proved in Russia in 1941–1944, did not eliminate air support, not only because the space was so great, but also because small units such as Stukas could survive on a battlefield in which air superiority was an hourly affair, not a continuous cloak.

In contrast to the usual defeat of air forces, the Battles of Britain and Malta stand out. Although Germany's ability to launch a successful invasion of England might be doubted, RAF Fighter Command was hard-pressed in the summer of 1940, and the Luftwaffe's change in focus from attacking its opponent's infrastructure to attacking the capital city had stunning consequences. In Malta an air service under siege and with few resources nevertheless held out for two years against great odds, thanks largely to the German failure to launch a combined assault on the island's airfields. For the RAF, both battles were earned victories to be sure, but not completely so.

Another atypical case is that of China in World War II, when the Chinese air force survived only because it withdrew to reorganize and re-arm before taking the offensive. It even went so far as to establish bases behind enemy lines, thanks to the spaciousness of the country and the weblike distribution of Japanese occupation forces along the Chinese railroad system. With a modern infrastructure that no partisan has ever enjoyed—fuel, ammunition, repair facilities—the Chinese air force conducted perhaps the closest thing we have ever seen to a guerrilla air war.

Neither of the post-1945 Chinese air forces, the People's Liberation Army Air Force (PLAAF) or the Taiwanese air force, has fought long enough to reach a decisive point. The trials and tribulations of these two Asian air arms have been long. But there has been no Far Eastern "Battle of Britain," for the PLAAF has never tried to overwhelm the Formosan Nationalist forces as a prelude to invasion. Perhaps this reflects the "paper dragon" nature of the Chinese—a people generally on the defensive rather than being a dangerous offensive power—or perhaps it represents the fact that mainland China has had too many political, social, economic, and industrial problems to become a Great Power.

Since 1947 the tensions between India and Pakistan have led to two wars and a series of confrontations. Neither side has won decisively.

In fact, it can be argued that Hindu India does not want to conquer Muslim Pakistan, and the latter's objective is simply to prevent Indian domination. Both sides have to face more than one way, and both have doubtful logistics, being vulnerable to the vagaries of outside suppliers.

Lessons of Defeat

By examining the defeats of air forces, can we discern any patterns, learn any lessons?

The dead ducks were doomed because they lacked the infrastructure and the wherewithal to withstand their fates, in large part because the foundations on which they were erected were destroyed. Their countries were occupied or disbanded. In most of these cases, smaller successor air forces emerged, but few had the resources to create a new aircraft industry, so they still had to rely on outside suppliers.

Of the new states, Poland created a vulnerable industry in the interwar years, but by 1939, the country was indefensible because it did not have the space or the manpower to oppose its powerful German and Russian enemies, nor was it backed by a viable alliance system after World War I. Moreover, the cost and demands of what was, by 1939, a modern air force were beyond Poland's means. In contrast, its enemies, Germany and Russia, were phoenixes that had the Mahanian necessities to rise again, albeit under dictatorships that recognized the potential of aviation and saw it through the technological revolution. In this process, politics and personalities played an important role, but so did space. The Soviet Union had enough space and the management know-how to move the industry beyond the Urals while still introducing new aircraft types. It used draconian measures to ensure output while concentrating on a few war-winning tactical types in a massive blitzkrieg. Whereas the USSR could trade space for time and, in the critical days of 1941–1943, was forced to fall back on its resources, such was not the case in Germany.

In Nazi hands, grand strategic blunders included the failure to anticipate the nature of the war in the air as well as on the ground and under the sea, the interference of politics, and inept management of the war economy. They also included the failure to envision the single-mindedness of the heirs of Giulio Douhet, Hugh Trenchard, and Billy Mitchell in developing in the hothouse of war the grand strategic bomber force, which shrank German space to the point where every industrial complex became vulnerable and the defense of the Third Reich involved a disproportionate share of man and woman power. War production began to decline in spite of the development of "wonder weapons," in part because, by 1944, Allied intelligence had the means to pinpoint the weaknesses of the economy, and in part because Hitler's enemies were advancing on the surface, supported by massive tactical air forces and an immense war economy in both the Soviet Union and invulnerable North America, whose spaces included training airfields and other vital establishments, as well as unlimited fuel.

Japan was strangled by maritime forces in the vast spaces of the Pacific and was gradually brought down by intra- and interservice and internal domestic politics, as well as by the same misallocation of resources to a future threat as occurred in Germany, leading to the loss of technical leadership. But due to Cold War necessities, it rose again as a self-defense force, supported by a necessary but not burdensome infrastructure.

The defeat of France was a result of a rotted political system and a military dominated by the marshals and their heirs, notably the chief of the army and defense staff. France fell behind technologically, but it also failed (partly for lack of credits) to provide the trained aircrews and mechanics to man the new modern combat aircraft finally being produced from 1938 on. Nor did it provide the necessary infrastructure to control airpower, and it was unable to foresee the multiple requirements of even an openly defensive war.

Germany, Japan, Italy, and France all benefited in the postwar world, but the most striking example of the phoenix is France. In a way, it was fortunate that it emerged from the war in 1944 with its aircraft industry completely stripped but with its technical personnel intact. This, coupled with a new political will that included consistent funding, not only produced excellent modern, exportable combat aircraft but also established the core of the Airbus family. This was not an isolated phenomenon, for it was boosted by the Cold War of 1947–1990, by NATO and the evolution of the European Community, and by the replacement of the Eurocentric, impotent League of Nations with the global United Nations, as well as by the enormous growth of air travel.

Of the other defeated air forces, the Egyptian benefited from the Cold War and grew to understand the necessity of overcoming its economic, political, and religious drawbacks. By 1973, it had created a new air force capable of countering the Israeli threat, thus engendering a stand-off peace not only with the chief enemy of the Arabs but also with its western neighbor, Libya. As with many other air forces, the Egyptian one is a client force dependent on external supply, although it is now becoming self-sufficient in manpower. Costs, of course, preclude lesser powers from creating a whole viable infrastructure.

The Cold War also saw the air forces that had been victorious in World War II, notably the U.S. Air Force and the Red Air Force, continue to stand on the cutting edge. But the different political systems enabled the Americans to become the world's leader into the twenty-first century, while the USSR's air force, with little postwar combat experience except vicariously through its satellites in the Middle East and Asia, crumpled when the state disintegrated in 1990, as Austria-Hungary had in 1918. The Soviet aircraft industry fell badly, only to be revived in small part by a drastic shift that made design bureaus responsible for production while at the same time compressing the former unemployment system into a capitalist one.

In terms of the aviation industry, the British have been forced to join with other European firms in partnership on international projects. Costs and demand required this, just as Boeing has had to share its work with both European and Asian partners and subcontractors. In all this, the RAF has seen its role shrinking and its infrastructure merging into the greater air community. The Cold War especially benefited the U.S. aircraft industry, which supplies both the U.S. Air Force and U.S. Navy, as well as allies and clients. The American air forces have been shrinking, however, as costs have risen and airframe life has been dramatically extended by electronic refits because of miniaturization and modernization—old airframes can do new things—fuel- and labor-saving devices, vastly enhanced controls, and the rise of the unmanned combat aircraft.

Whether defeated or not, air forces face the weakening complexities of costs and controls and the consequent shrinking of their size. As enmities end, the civilian, commercial side expands. Whether or not defeat is absolute, the consequences are closely related to the patterns of politics and development.

Recommended Reading

Some good places to start are Peter Paret and Gordon A. Craig, eds., *The Makers of Modern Strategy from Machiavelli to the Nuclear Age* (Princeton, N.J.: Princeton University Press, 1986); Alfred Thayer Mahan, *The Influence of Sea Power on History, 1660–1783* (Boston: Little Brown, 1890); and Williamson Murray, MacGregor Knox, and Alvin Bernstein, eds., *The Making of Strategy: Rulers, States, and War* (Cambridge: Cambridge University Press, 1994). See also Enzo Angelucci, *The Illustrated Encyclopedia of Military Aircraft, 1914 to the Present* (Edison, N.J.: Chartwell Books, 2001); Andrew H. Cordesman and Abraham R. Wagner, *The Lessons of Modern War* (Boulder, Colo.: Westview Press, 1996); M. J. Armitage and R. A. Mason, *Air Power in the Nuclear Age* (Urbana: University of Illinois Press, 1983); Robin Higham, *100 Years of Air Power and Aviation* (College Station: Texas A&M University Press, 2003); Tony Mason, *Air Power: A Centennial Appraisal* (London: Brassey's, 1994); John Buckley, *Air Power in the Age of Total War* (Bloomington: Indiana University Press, 1999); and Alan Stephens, ed., *The War in the Air 1914–1994* (Canberra: RAAF Air Power Studies, 1994).

Histories of individual air forces are rare. There is no scholarly study of the Royal Air Force, the U.S. Air Force, the Luftwaffe (except for the Air Ministry account: *The Rise and Fall of the German Air Force, 1933–*

1945 [New York: St. Martin's Press, 1983]), or the Japanese air force, at least in English. A two-volume study of the French Armée de l'Air is nearing completion in 2005. The Red Air Force is the subject of Robin Higham, John Greenwood, and Von Hardesty's *Russian Aviation and Air Power in the Twentieth Century* (London: Frank Cass, 1998). There is also an official work on the People's Republic of China: Duan Zijun, *China Today: The Aviation Industry* (Beijing: China Aviation Industry Press, 1989). See also Victor Flintham, *Air Wars and Aircraft: A Detailed Record of Air Combat, 1945 to the Present* (New York: Facts on File, 1990).

Some specialized works should be consulted for more specific overviews, such as Herschel Smith, *A History of Aircraft Piston Engines* (Manhattan, Kans.: Sunflower University Press, 1986); Peter L. Gray and Owen G. Thetford, *German Aircraft of the First World War* (London: Putnam, 1962); William Green, *Warplanes of the Third Reich* (New York: Doubleday, 1970); Ray Wagner, *American Combat Planes* (New York: Doubleday, 1982); and Rene J. Francillon, *Japanese Aircraft of the Pacific War* (London: Putnam, 1979). See also Emmanuel Chadeau, *L'industrie aéronautique en France, 1900–1950* (Paris: Fayard: 1987), Benjamin Franklin Cooling, ed., *Case Studies in the Achievement of Air Superiority* (Washington, D.C.: Center for Air Force History, 1994); and Stephen Budiansky, *Air Power: The Men, Machines, and Ideas that Revolutionized War, from Kitty Hawk to Gulf War II* (New York: Viking, 2004).

CHAPTER ONE

Poland's Military Aviation, September 1939

It Never Had a Chance

Michael Alfred Peszke

The short interwar (1918–1939) history of Poland's Military Aviation (Lotnictwo Wojskowe) is a paradigm for the history of Poland's efforts to ensure its security. The problems that confronted the Poles were two disgruntled neighbors, Germany and the Soviet Union, unhappy with territorial losses and seeking revindication; close to indefensible boundaries, particularly those with Germany; and a disastrous economic situation.

Poland embarked on independence in 1918 with industrial output at 20 percent of its 1913 production. This was primarily due to the fact that partitioned Poland had been the battleground of World War I military operations between the Russians and the Central powers. The loss of Polish industry in that period was estimated at 73 billion French francs. Furthermore, the worldwide crisis of 1929 hit Poland severely, with every fourth Polish worker being unemployed. Poland's per capita annual income was 610 zlotys, compared with the Western average of 2,490.

The military disaster that the Poles suffered in September 1939 is well characterized by the words of German historian Horst Rohde: "The inadequacy of Polish political judgment was reflected in the belief of effective support from Britain and France and in the neutrality of the Soviet Union." Although these comments are true in retrospect and reflected the reality of September 1939, Poland's interwar foreign policy was based on nonaggression treaties with its neighbors, reinforced by concrete military alliances with the French (February 1921) and Romanians. The most assiduously sought alliance—with the British—was partially achieved in March 1939 and solidified in August by a treaty of mutual assistance. The British government had agonized about the choice between the Soviets and the Poles for its eastern "front" and finally assessed the situation as follows: "it was better to fight with Poland as an ally than without her."

Historical Provenance

Poland's Military Aviation in 1919 was an ad hoc creation, and most Polish pilots were veterans of the First World War who had flown for other countries. The Polish fliers were augmented by a significant number of French and Italian pilots and even some Americans who formed the Kosciuszko Squadron. They flew a motley collection of abandoned German planes, surplus aircraft from the French and Italians, and even a gift from King George V of Britain. But the various Polish squadrons performed a heroic task of providing valuable reconnaissance and even support to the Polish ground forces during the border war with Russia. The ground commanders were enchanted by the depth added to their defensive and offensive capabilities by the aviation service, primitive as it was. They wanted more of the same. At the war's end in 1920, there were a total of sixty serviceable aircraft, but once the French, Italians, and Americans went home, the number of skilled Polish personnel was limited. With Polish–French relations being especially cordial at this time, the French provided a loan for the purchase of military equipment, and French general Francois-Leon Leveque was appointed to command the Polish aviation service.

Early Operational Doctrine

The aviation service was a constituent element of the army, with an identical administrative structure; the six aviation regiments, like infantry or artillery regiments, were major administrative bases, responsible to the local army corps commanders and ultimately to the minister of military affairs.

The aviation regiments were responsible for the maintenance of equipment, the induction and training of conscripts, and general quartermaster services. Each regiment had a mixture of fighter, army cooperation, light or medium bomber, and tactical support aircraft. For a short time, the all-fighter regiment at Lida was an exception to this general rule.

Leveque's plan called for 575 aircraft divided into eight "line" wings, eight fighter wings, and one bomber wing. Using 400 million francs of the French loan, the Poles purchased French Potez 15s, Hanriot HD 14s, and four flying boats and procured licenses for the construction of some French aircraft in Poland. Leveque's goal was to make the air force an auxiliary weapon cooperating closely with the army, and he was the author of the first thorough regulations defining the mission of the service.

General Zagorski replaced Leveque and undertook an ambitious development program. New air bases were created, existing ones were improved, and underground storage depots were constructed. It was at this time that the aviation service's officer academy was initiated (corresponding to the academies for the infantry, artillery, and other service specialties); it was eventually based at Deblin and was known proudly all over Poland as the "Eagle's Nest." Under Zagorski, the Polish air industry received licenses to build French planes. To initiate this buildup, Zagorski purchased 600 mostly obsolescent planes from France, including 300 Spads 51 and 61, 250 Breguets, and 32 Farman Goliath bombers, for which there were no crews or hangars.

Zagorski's plan (never realized) called for a force four times greater than that originally planned by Leveque. Poland's Military Aviation was to consist of five types of units: army cooperation, bomber, fighter, pursuit, and tactical (i.e., bomber-reconnaissance). Bomber and pursuit squadrons were to be centralized under the command of the commander in chief; the rest would be allocated for army cooperation and support. The commitment of so much capital for the acquisition of planes for which there were insufficient trained personnel (the existing seventeen squadrons had only about half the necessary officers and pilots), no hangars, and inadequate operational doctrine did not find support in the Polish General Staff.

Zagorski resigned in March 1926. He was a controversial figure, admired by some as an aviation visionary and condemned by others for his alleged close ties to the scandal-ridden French venture company Francopol, which served to enrich many French contractors. Francopol was liquidated in 1927 after having produced only two engines.

Colonel Ludomil Rayski assumed command of the aviation service and single-mindedly pursued the development of a Polish aircraft industry to ensure self-sufficiency. But because the prior contractual agreements with the French had to be honored, the aviation service continued to be the recipient of obsolescent and badly designed French planes. In the words of Belcarz and Peczkowski, "flying Spads threatened rapid annihilation of the entire (Polish) fighter pilot force."

In 1926 General Joseph Pilsudski staged a coup. At the time, there were both internal domestic problems and a deterioration of the international situation owing to the Treaty of Locarno (1925), which regularized German western boundaries but failed to do so for the east. Poland felt betrayed and protested, but to no avail; the French alliance was crippled, and trust was impaired.

Dual Command Structure

Pilsudski established a new dual command structure. The general administration of the military (and thus the aviation service) was under the minister of military affairs, who was a member of the cabinet. The new post of inspector general was also created, a post unencumbered by administrative issues and confined to being a planning staff for war. A number of senior generals (inspectors) reporting to the inspector general were responsible for assessing whether the units and their equipment met the necessary criteria. The theoretical underpinning for this dual command structure was that the inspectorate proposed, while the ministry executed. In retrospect, it is impossible to find any redeeming features in this arrangement. One can only speculate that it was a political move to allow Pilsudski to control the military without ministerial accountability, as he held both posts until his death.

In 1926 Pilsudski issued his one and only order regarding the aviation service: it was to address the excess of planes and the inadequate personnel and garrisons and to confine its activities to reconnaissance and communications. Thus, in the continued polemics, that order reflected either his ultimate ignorance or his pragmatic recognition of Polish reality. The debate continues.

Polish Aviation Industry

Most of the Polish military industry, including aircraft production, was nationalized by 1935. The major aircraft manufacturer, Panstwowe Zaklady Lotnicze (PZL), became state owned. This coincided with Rayski's policy of equipping the service with Polish-made planes—a policy that, on the surface, progressed in a satisfactory manner. The Polish fighter squadrons were the first in the world to be equipped with all-metal monoplanes designed and built in Poland. However, all were powered by foreign (British Bristol Company) engines, most of which were built in Poland under license. The crucial factor that hampered an indigenous aviation industry continued to be economic. Engineers had to be recruited and trained, and to ensure an adequate supply of young, skilled professionals, polytechnic departments had to be initiated. The new sciences of metallurgy, precision lathe machining, and engine manufacturing all had to be started from scratch. There was a major shortage of investment capital in Poland, and the military constantly competed with other state needs for its share of inflation-ridden and economically depressed budgets. Still, by 1939, the eight aircraft manufacturing factories employed 12,400 workers and had built 1,127 planes under license and 2,458 of Polish design; they had produced 3,550 engines, but only 150 of strictly Polish design.

Evolving Operational Doctrine

In addition to promoting an indigenous aviation industry, the operational doctrine of the service was undergoing continual evaluation and evolution. The army corps commanders (accountable to the minister of military affairs) and army inspectors (accountable to the inspector general) had formal administrative authority and inspection jurisdiction, respectively, over the aviation regiments in their districts. Some were disinterested and left Rayski to manage as best as he could, but there were exceptions. In 1927 army inspector Edward Smigly-Rydz (in 1939, he would be commander in chief of the Polish forces in the September campaign) expressed his views that aviation planning could not extend beyond three years because of the rapidity of technological changes and that more attention needed to be given to fighter defenses. He was also instrumental in calling for an active review of the aviation service regulations. In 1929 the General Staff formed an expert commission to address specific aviation problems; its members included two senior aviation officers, Stanislaw Kuzminski and Stanislaw Ujejski. This commission recommended that, due to its small size, the aviation service be centralized and equipped with multipurpose aircraft. Shortly thereafter a second, larger commission was convened that included a number of distinguished aviation officers and two famous Polish aviation theoreticians—Abzoltowski and Romeyko. The group worked for three years before its report—the Regulamin Lotnictwa (aviation regulations)—was finally completed in 1931. This was the first doctrine that spelled out the need for supporting services and called for four types of aircraft: bomber, fighter, tactical, and army cooperation (at the time, the concept of pursuit squadrons was omitted but was added later in the decade). Bomber squadrons were to be equipped with the LWS 4 Zubr, and fighters with the continually updated PZL 11. Tactical squadrons were to be equipped with planes providing tactical reconnaissance for army commanders with some capability for bombing; this led to specifications that gave rise to the PZL 23 Karas. Army cooperation aircraft were to be the proverbial dogsbodies, and those specifications eventually led to the LWS Mewa.

These regulations, with a number of relatively minor modifications, were the operational doctrine under which the Poles went to war in September 1939. But it should be noted that Rayski was on record as urging the integration of the command of all military aviation and antiaircraft defenses. This was eventually implemented in 1937, but the position was given to General Jozef Zajac. Rayski also had his own vision of the importance of his service, which he saw through the fashionable prism of the Douhetian doctrine. So as early as the late 1920s, there were seeds of potential discord between the staff and its commissions, reflecting a rather conservative point of view in which the aviation service's role was seen as auxiliary, and Rayski, who wanted a strategic bomber strike force.

In 1936 Polish Military Aviation was very much a product of Rayski's procurement policies and consisted of 318 planes in thirty-three squadrons: seventeen light bomber-reconnaissance squadrons (equipped with Polish-built Potez 25s and Breguet 19s, soon to be replaced by the Polish PZL 23 Karas), thirteen fighter squadrons (PZL 7s and 11s), and three bomber squadrons (Polish-built Fokker FVIIs, eventually replaced by the unsuccessful LWS 4 Zubr.) In addition, there were thirty-three flights of army cooperation planes with little combat potential (R-XIIIs, Czaplas, and RWD 8s.) The big Goliaths were never entered into combat service and were left out in the open in all kinds of weather. Some were used for parachute training.

Changes in Command Structure

In many ways, 1936 was a watershed year. After Pilsudski's death, General Smigly-Rydz assumed the mantle of inspector general, and General Tadeusz Kasprzycki became the minister of military affairs.

In the mid-1930s there were signs of major change in the aviation service. The uniform color was changed to steel blue, beginning the process of building up its own identity. An aviation staff academy was formed, and in 1937 an aviation service staff was created as part of the General Staff of the Inspector General. A seminal and, ultimately, destructive step was the creation of an inspector of aviation and antiaircraft defenses, which resulted in the aviation service being led by three independent planning nuclei.

These three authorities, largely autonomous and even competing, were directed, respectively, by the commanding officer of the service (Rayski), reporting to the minister of military affairs; the inspector of the aviation service (Zajac, appointed 5 January 1937), reporting to the inspector general; and the head of the aviation staff (Ujejski), which was part of the General Staff. The formal relationships among these positions were not clarified and had egregious shortcomings, since the inspector of aviation (the wartime general officer, commanding) had neither a staff nor procurement authority and did not participate in meetings of the aviation staff. General Zajac, in addition to being inspector of military aviation and antiaircraft defenses, was the commanding officer of all antiaircraft defenses, which placed all planning and command functions with one individual.

Germany's 1936 entry into the so-called demilitarized region west of the Rhine, contrary to the Treaty of Versailles, led to alarm in France and an almost immediate invitation to Smigly-Rydz to visit France. This breathed new life into the old alliance, and Poland obtained a significant four-year loan for military industrial development and for the purchase of military material in France. This loan led to the creation of the Centralny Okreg Przemyslowy, sited in central Poland, and to the reactivation of a standing committee, the Komitet do Spraw Uzbrojenia i Sprzetu (KSUS), chaired by General Sosnkowski, whose function was to allocate the funds among the various competing demands.

In 1936, as part of the rethinking of Polish strategic plans, a section of the General Staff submitted a plan for the operational use of the aviation service. It was based on the experiences of the annual air maneuvers and the Douhetian theory that a bomber would always get through and that a powerful bomber force could determine the outcome of any war. The plan called for a bomber force sufficient to interdict all German military movements east of the Oder River. Such a strategic intervention would require a bomber force of 378 planes in sixty-three squadrons augmented by 360 tactical planes (the old concept of bomber-reconnaissance) modified for dive-bombing.

Concurrently, Rayski proposed a plan (which would have been completed by 1942) calling for a total of 886 planes broken down as follows: a central independent force under the commander in chief of thirty bomber and thirty-two tactical squadrons, and a tactical force assigned to field army commands of eight fighter, eight reconnaissance, and eighteen army cooperation squadrons. The cost of such a plan was estimated at 1.153 billion zlotys, exceeding the entire amount of the four-year loan being negotiated with France. On 7 July 1936 Rayski was advised that the costs were prohibitive, and KSUS scaled down the growth of the service to 252 bombers and 336 tactical dive-bombers.

Rayski persisted with his strategic goal but agreed to reduce the total bomber strength from 886 to 708 (compared with the KSUS proposal for 252 bombers), reallocating the money saved by reducing the number of fighter and tactical squadrons. Thus the aviation budget was reduced to 1 billion zlotys to be spent over the next eight years. This amounted to 20 percent of the whole rearmament budget, which came from the following sources: French loan, state bonds, voluntary contributions, and the export of Polish-produced military hardware (which in 1938 alone came to 186 million zlotys). The sale of bonds was also successful, particularly the antiaircraft defense bonds, which were intended to raise 100 million zlotys and actually realized 390 million.

Aviation Maneuvers and War Games

In the 1930s Poland's aviation service participated in a number of major maneuvers. In 1934 a total of 232 planes took part in exercises that addressed the problem of the movement of air units and their supporting ground units and the necessary combat concentration of planes in the air. A similar exercise took place in 1936, again with more than 200 planes participating.

In 1937 the antiaircraft defenses staged an exercise postulating the defense of Lodz, with the participation of officers from the aviation staff academy and communications, antiaircraft artillery, and military rail and balloon sections. The blue (Polish) side consisted of more than one hundred officers, and the red (German) side of thirty-five. Also in 1937 the Karas bombers of the Warsaw Aviation Regiment carried out a demonstration attack against a rail junction, causing a communications disruption lasting more than twenty-four hours. The head of military rail communications took great pains to address such damage.

In September 1938 (the traditional month for military maneuvers in Poland, after the harvest) nearly 200 planes, antiaircraft defenses, and even the embryonic airborne units carried out exercises to synchronize the tactical use of the aviation service and the active and passive antiaircraft defenses.

Preparations for War

General Zajac devoted his analytical skills to the prospect of conflict with Germany and came to the conclusion (which has been severely condemned by ad hominem judgments) that the Rayski plan for a war-winning bomber force was not feasible. Zajac pointed out that the planes in development or production would enter service in sufficient numbers only in late 1940 or, more realistically, in 1941. In his sober and detailed memoirs, Zajac writes that in the summer of 1938, after visiting both France and the United Kingdom, he concluded that Poland could not expect any material help from these Western democracies. He also opined that in the event of war with Germany, Poland's aviation service could survive only four weeks at best. This analysis led to a detailed report, which led to changes in aircraft procurement that continue to be controversial to this day.

Zajac concluded that Poland's primary air weaknesses were its inadequate system of communications and radio navigation and the outdated equipment of its fighter squadrons. The first two precluded a successful bomber offensive. Zajac used his authority to develop plans to correct the inadequacy of the fighter forces and, stressing the emergency nature of this situation, urged that foreign planes be purchased. This was a philosophical departure from Rayski's policy of self-sufficiency and also from his concentration on a bomber force.

It is pertinent to point out that in 1936 General Tadeusz Kutrzeba, who was commandant of the Polish War/Staff Academy, analyzed the

Polish military situation and reported to the inspector general that in the event of a German attack, Polish forces could resist for only six weeks unless France made a concerted effort in the west. This evaluation most likely would have held up had the Soviets not attacked Poland in the east on 17 September 1939.

It is interesting that at approximately the same time, the British Royal Air Force (RAF) also changed its basic operational doctrine and began to emphasize fighter forces. Scheme J in October 1937 called for a U.K.-based strike force of 1,442 planes, of which only 532 were fighters; by January 1938 the plan was scaled back to 1,369 bombers, and in October the RAF planned on a significant increase in fighter defenses to 800.

In March 1939, on learning that General Zajac would be the aviation commander in the event of war, Rayski offered his resignation and became a deputy minister of military affairs. Once the staff had accepted Zajac's vision of the aviation service, Rayski's departure was inevitable.

General Kalkus succeeded Rayski as commandant of the peacetime aviation service, and in March 1939 he received orders to put the aviation service on a war footing. Kalkus was empowered to use the 1939–1940 budget to acquire war materials and auxiliary equipment and to purchase tactical and fighter planes, as well as to requisition the advanced Karas that had been built for Bulgaria. Between March 1939 and the onset of war, significant logistical systems were built up. Poland was divided into the war zone, where the combat armies would be deployed (territories west of the line running from Cracow to Torun), and the rest of the country, which would be the administrative and supply zone under the authority of the minister of military affairs. Kalkus was also ordered to evacuate bases from the exposed western territories to central or eastern Poland. As a result, the Cracow air base was moved to Lwow; those at Poznan, Torun, and Bydgoszcz were evacuated to Lublin and Malaszewicze (near Brzesc). No plans were made to evacuate Warsaw or Deblin because nobody thought the situation would get that bad.

Kalkus also stockpiled war material according to the following principles. All combat units were to have a seven-day supply of gasoline, oil, and bombs. The aviation commander (Zajac) would have another ten-day supply stockpiled and deployed in various rail junctions west of the Vistula; these supplies could be moved forward or back. There would be another three-month supply of gasoline in the main aviation bunkers at places such as Pulawy (near Deblin), Lublin, and Malaszewicze.

Great effort was also focused on constructing a system of airfield clusters in areas where aviation units were to be deployed. These clusters included secret forward fields, auxiliary fields, and even simulated fields with mock planes. By 1939 there were 38 such secret operational fields and a total of 221 auxiliary fields that served the needs of the aviation service well, except for the twin-engine PZL 37 Los bomber, which could not take off from there when fully loaded.

Zajac's views had prevailed, but too late. Poland's air force was, in effect, caught changing horses in midstream. Orders for the Los were cut, and orders for the PZL 50 Jastrzab fighter plane were put on hold due to its unsatisfactory performance. In these last days of peace, the Polish authorities were confronted with the reality that neither the single-engine fighter nor the twin-engine pursuit plane could be produced. The decision to suspend orders for the excellent Los has been severely criticized and cited as evidence of the Polish military staff's incompetence. The reality is that Poland did not have enough resources to build up both its fighter and its bomber forces, and in hindsight, its goal of having a bomber strike force was a fantasy. This strategy did not work even for the British, whose insular bases remained relatively safe throughout the war. Rayski erred, as did many other aviation enthusiasts of that period, whereas Zajac was right but lacked the time to correct the problem.

France was initially reluctant to sell its Morane-Saulniers to Poland. Finally, in late summer, a contract for 140 of the French fighters was signed, with deliveries to begin in early fall. The Polish air staff considered a number of American fighter planes—those manufactured by Brewster, Curtiss, Grumman, and Seversky. The most serious negotiations were with Curtiss, since the French had already purchased this plane, giving it their stamp of approval. The price was $8 million for 143 planes. Allegedly the American middleman offered a loan with 3 percent interest to facilitate the sale. On 21 July 1939, however, the Polish military attaché in Washington was advised that the deal was off due to the lack of hard currency.

The British were not prepared to sell their Spitfires, but on 16 June 1939 the Polish air attaché in London cabled Warsaw, requesting that four experienced pilots be sent to the United Kingdom, since there was some break in the negotiations for the sale of fourteen Hurricanes. The British posture was explained by the small British credits available for Polish rearmament. Still, in July 1939, Minister of Military Affairs Kasprzycki advised General Zajac that twenty Fairey Battles would be arriving from the United Kingdom every month until this order was filled. But the Fairey Battle was already obsolescent, even though it equipped frontline RAF squadrons through 1940, including the first two Polish bomber squadrons (300 and 301) formed in the United Kingdom.

Although the quality and quantity of equipment had degenerated to a critical point, the training of personnel had reached a satisfactory level. From the fall of 1937 to the fall of 1938 the number of trainees reached 2,500, including 730 officer cadets in Deblin and more than 1,000 noncommissioned officers at Krosno and Bydgoszcz. There were also special courses for armorers, mechanics, radio specialists, and truck drivers (not a common skill in prewar Poland). In addition, about 400 noncommissioned pilots and air gunners, who were mostly reservists, were called up by the secret mobilization orders issued on 25 August 1939. The goal was to increase the number of cadets at Deblin to 1,100. In late 1938 and early 1939 this ambitious plan began to pay off; the air and ground crews were now well trained and had reached near 40 percent reserve capacity.

Plans for the further growth of the service continued, and decisions were made to assign dispositional air units to army commands. In the year prior to the outbreak of the war, the aviation section of the General Staff, headed by General Stanislaw Ujejski (inspector of the

Polish air force in the United Kingdom between 1940 and 1943), took the initiative and assumed responsibility for the development of operational doctrine.

In eastern Poland, far from Germany, a new base at Brzesc was created. The existing air regiments were renamed air bases, and the name regiment was given to the squadrons assigned to the army (or tactical units), of which there were to be six, including fighter, reconnaissance, and army cooperation. The dispositional air force was to have one bomber brigade based at Brzesc; two tactical brigades at Skierniewice and Krasnik; and three pursuit brigades at Warsaw, Bialystok, and Ostrowiec. The philosophy behind these new plans was to give the army air regiments full mobility, while the dispositional air force would be serviced by the permanent bases and restricted for use by the commander in chief.

During the summer of 1939 politicians and diplomats of the Western democracies (Poland's allies) strove mightily and with increasing guile (if not duplicity) to prevent war. The Poles were clearly concerned that the final result of this peace-seeking process would be a new version of Munich, with Poland being asked to pay the price of "peace in our time." But a number of military staff meetings between the Poles and their ambivalent Western allies did take place.

The first of these occurred in Paris on 16 May 1939, with the Poles represented by Minister Kasprzycki and the French by Generals Gamelin and Georges. The French representatives expressed concern that the primary attack could be an invasion of France by combined German and Italian forces and thus expressed the hope that Poland would attack Germany. In the event of a German attack on Poland, Gamelin assured the Poles that the French army would develop an offensive by the fifteenth day of the war, and air operations by the French would be initiated immediately. This meeting led to the formation of a special subcommittee on air matters, with the Poles represented by Colonel Karpinski of Poland's Military Aviation and the French by General Vuillemin, chief of the French air force General Staff. The French agreed in principle to move five bomber squadrons to Polish bases for the shuttle bombing of German targets. In fact, preliminary steps in this direction were implemented. On 8 August 1939 a French Amiot E 358 carrying staff officers landed at Warsaw's Okecie airport to explore the feasibility of such operations. The Polish air staff mobilized a small group of experts to work with the French and prepared the landing fields in western Poland. The joint Polish–French commission agreed to undertake immediate negotiations on all the essentials involved with such plans, such as radio communications, logistical support, and the preparation and maintenance of bases on Polish territory. Gamelin made the military agreement conditional on the political convention, which the French dragged their feet on until 4 September, the day after France declared war on Germany in tandem with the British.

On 19 July the British chief of the Imperial General Staff, General Sir Edmund Ironside, arrived in Warsaw and held a number of meetings where he was forthcoming in promising aid but made no commitment in terms of direct British military action. In diaries published after the war, Ironside commented, "The French had lied to the Poles in saying they are going to attack. There is no idea of it." Also during these staff talks, the Poles gave a copy of a reconstructed Enigma machine to both their French and British allies.

In spite of protracted negotiations, the British treasury was loath to grant Poland a substantial loan for the purchase of military equipment, especially from the United States, where such equipment was available. But it grudgingly agreed to extend credit for purchases in the United Kingdom, where the Poles were advised that modern planes were not available for export. The British Foreign Office supported the Polish request for financial assistance, and Ironside cabled London from Warsaw in July 1939 that the Polish effort to prepare for war was "little short of prodigious" and added, "we ought not to make so many conditions to our financial aid. That one of the ways of convincing Hitler that we are serious is by granting this monetary aid to Poland. That the Poles are strong enough to resist."

For Poland, the impending war was to be a joint effort with its allies, and the disposition of the Polish army, which has often been criticized, has a logical explanation when both Polish military strategy and

Polish political fears are taken into consideration. Poland's strategy was basically to be the morass into which German armies would be drawn while the French attacked in the west. Thus the divisions were positioned close to the border, with the simple intention that the Germans would have to develop their resources each time a defensive point was reached. The political concern was that if the Germans occupied parts of western Poland (which had been in contention since the 1919 Treaty of Versailles) and then asked for arbitration, the Western democracies might be tempted to pursue the Munich road of appeasement.

Poland's Military Aviation went to war equipped largely with early 1930s planes, with the exception of the PZL 37 Los, and with an ad hoc division of its forces into army (tactical) and dispositional (strategic) squadrons. Some Karas squadrons were retrained for bombing missions and became part of the Bomber Brigade, along with the two Los wings. The actual strength of the Polish Military Aviation on 1 September 1939 was nine bomber squadrons (86 planes), which were part of the Bomber Brigade directly under the commander in chief, and the five fighter squadrons (50 planes) that composed the Fighter Brigade, whose task was the defense of Warsaw. Army commanders were assigned ten fighter, seven tactical, and eleven army cooperation squadrons, for a total of 274 planes. There were 16,000 officers and men, of whom 1,181 were pilots, 497 navigators, and 219 air gunners.

In August 1939 Inspector General Smigly-Rydz issued an order to all aviation commands restricting bombing to strictly military targets. It spelled out the various parameters and definitions but concluded that the spirit rather than the letter of the law should guide a conservative approach. On 25 August secret mobilization orders were issued for ground components to be moved to secret bases. On 31 August the majority of the air combat units were flown to these bases. The implementation of this order is beyond any historical doubt. German historian Cajus Bekker has gone on record as stating, "Despite all assertions to the contrary, the Polish air force was not destroyed on the ground in the first two days of fighting. The bomber brigade in particular continued to make determined attacks on the German forces up to September 16th."

War

The German attack on 1 September destroyed many civilian, training, and reserve planes, and the heavy bombing of the Deblin region inflicted severe damage on the logistical support system of the Polish aviation forces. Therefore, it is inaccurate to describe the intensity of the Luftwaffe's attack as inconsequential, just as it is erroneous to credit it with a one-day victorious stroke The performance of the various Polish aviation commands varied considerably. Though clearly outperformed and outnumbered, many squadrons continued to be in the fight. Problems arose from disrupted communications and when bases were overrun by the fast-advancing German motorized units.

Fighter Squadrons

There were 158 available fighter planes, mostly the PZL 11c but also 30 of the PZL 7a. The Fighter Brigade (commanded by Colonel Pawlikowski, killed in 1943 while leading a Polish fighter wing sweep over the English Channel) began the September campaign defending Warsaw. A well-organized alert system developed by Major E. Wyrwicki, the chief of staff of the brigade (killed in 1940 during the French campaign), depended on ground observers alerting a central control point, which in turn was in radio communication with flight commanders on patrol. The flights were then given ground coordinates to intercept German air attacks. In the initial period the brigade had forty-three victories.

As the forward observation posts were overrun, a decision was made to move the brigade southeast to the Lublin region. At that point the brigade, though reinforced by a number of fighter wings withdrawn from their armies, experienced administrative chaos. Short of gasoline and ammunition, the fighter planes wasted time. The wings of the Fighter Brigade were ordered to protect the bridges over the Vistula and to carry out shallow reconnaissance for the Lublin army, which was being desperately formed east of the Vistula. Only seven air victories were achieved in the last ten days before the brigade was ordered to Romania.

The fighter squadrons assigned to the army also had their heroic moments. Lieutenant Gnys of the Cracow army aviation wing shot down the first German plane of World War II. This wing (III/2) scored ten victories in the first three days of the war. In a series of sweeps, the fighter wing of the Pomorze army (III/4) claimed nine air victories on 2 September (Lieutenant Skalski, Poland's top-scoring fighter ace, downed two DO-17s). In the first three days, the wing achieved a score of eleven victories and continued to be effective even after moving east to Lublin, where ten more victories were achieved.

The performance of the Poznan fighter wing (III/3) deserves special mention. Under the command of the indomitable Major Mumler (later the commanding officer of No. 302 Polish Squadron during the Battle of Britain), the Poznaniaks stayed with their army colleagues on the ground until 16 September—longer than any other aviation unit initially assigned to an army tactical role. Their secret bases were not reconnoitered by the Germans, and they operated in an area that had been richly stocked with supplies in preparation for the French shuttle bombers. The number of air victories achieved by the Poznan wing was also impressive. In the first three days, the wing scored ten victories; in the next five days, it had seven more; and then between 10 and 17 September, it posted an additional fourteen—by far the best record. When it moved to Romania, it still had three fighter planes in operation. Altogether, Polish fighters shot down 129 German planes and lost 116, with 50 being evacuated to Romania on 17 and 18 September 1939.

Bomber and Tactical Squadrons

The Bomber Brigade, consisting of eighty-six planes, was formed from four wings and one squadron originating from a number of different bases. The two Los wings (thirty-six planes) were from the Warsaw base; the fifty Karas were from Cracow (II), Lwow (VI), and one squadron from Lida. This was the commander in chief's strategic bombing reserve, and even brigade commander Colonel Wladyslaw Heller could not undertake operations using more than one wing at a time. This limitation, undoubtedly meant to conserve resources and ensure flexibility in executing missions, proved to be a major handicap. Orders from the commander in chief were based on situational assessments from the army commanders, but these reports were often quite late in arriving, and by the time they were conveyed to the combat wings, the situation was often drastically different. Furthermore, the initial cluster of bases to which the Los bombers were dispatched proved to be unsatisfactory, and considerable time was lost as the wings were moved around.

When the Cracow 24th Army Squadron identified German armor cutting through the Polish lines north of Czestochowa, it carried out the initial bombing; the Bomber Brigade was then ordered to interdict the German attack. In the late afternoon of 2 September the Karas VI wing carried out a nineteen-plane bombing sortie. On 3 September the Bomber Brigade repeated the attacks with the Karas II wing, which carried out two missions involving a total of thirty-three planes. The Karas wings experienced heavy losses, and even the planes that made it back to base were badly damaged. These attacks directed against General Rundstedt's army group continued until 6 September. The Los bombers entered the war in full force only on 4 September, when more than forty sorties were flown. There was then a lull for two days as the wings moved.

At this point, the Poles became concerned about the pincer movement stemming from East Prussia commanded by General Bock. On 7 and 8 September more than fifty bombing sorties were flown against targets north of Warsaw. The lion's share of sorties on the eighth went to the Karas VI wing, which had only eight serviceable planes; it thus flew one- or two-plane missions, for a total of twenty-two sorties, from the early morning to dusk using its extra crews. By 10 September the losses and wastage (none of the Polish planes was armored) led to the disbanding of the II wing, which gave up its planes to the VI wing and was trucked to the Romanian border, expecting to receive the British Fairey Battles. At best, these British planes would have arrived in late September through Tulcea or Galatz, but it boggles the mind how the authorities expected the crews and mechanics to gain immediate familiarity with a completely different plane.

The communications and logistical system began to crack under the constant stress of moving from base to base as the Luftwaffe reconnoitered Polish airfields. The bomber wings went from being an effective strike force to being mere pinpricks, but their reconnaissance was of continued value to the Polish high command. By 8 September the bomber wings were at half their original strength. All told, during the seventeen days of combat operations, the brigade received ten Karas and nine Los replacements. It lost eighty-two aircraft (81 percent of its initial strength) and suffered human losses of 108 aircrew killed, wounded, or missing.

The Los bombers flew more than a hundred missions—most of them bombing, but some solely reconnaissance. The Los had a maximum bomb-load capacity of 5,688 pounds carried internally, but while operating out of the relatively primitive grass fields, it carried only half the maximum bomb load. This gives a plausible estimate that the Los bombers delivered at least 250,000 pounds of bombs. The more numerous Karas flew 130 sorties, but its maximum capacity (carried externally) was only 1,543 pounds, and in combat operations, it was limited to 800 pounds. Thus, it can be surmised that about 350,000 pounds of bombs were delivered by the Bomber Brigade on German troops.

When the Soviets invaded Poland in the early hours of 17 September, the remaining units, having lost all their bases, were ordered to evacuate to Romania. The last operating unit of Poland's Bomber Brigade, the Los X wing based at Buczacz, was bombed by Soviet planes on the afternoon of 17 September. The Bomber Brigade lost all its Karas bombers and twenty-six of its Los bombers.

The Karas squadrons assigned to the various armies suffered severe losses. Like the tactical fighter wings assigned to the armies, the Karas squadrons were pulled back to reinforce the Bomber Brigade. By 6 September they represented little combat potential. The exception to this was the 24th Squadron assigned to the Cracow army. It had taken part in the major bomber strike against German armor near Czestochowa and continued, with no material loss, to provide reconnaissance for the army. When pulled back due to the loss of its initial bases, it was assigned to the commander in chief for "strategic" reconnaissance. Commanded by Captain Julian Wojda (later a flight commander of No. 304 Bomber Squadron in the United Kingdom and Polish liaison officer to RAF No. 1 Bomber Group), this squadron operated right up to 17 September, remaining undiscovered by the Luftwaffe and managing to cope with logistical issues. When it was finally forced to evacuate to Romania, eight of its ten original Karas planes were still operational.

In the early hours of 17 September, the Soviets entered Poland, and the aviation units—which had been based as far east as possible, away from the German encirclement—were directly exposed. Smigly-Rydz ordered all military units that could do so to cross the Romanian and Hungarian borders. The aviation units were already in a favorable position for this denouement; they had organic transport based close to Romania and organic motor transport, as well as many serviceable planes. But the ground forces continued to fight in central Poland, around Warsaw, and in the Hel peninsula.

By October 1939, 9,276 personnel of Poland's Military Aviation had crossed into Romania; about 140 Polish planes had also flown into internment, including 54 fighters, 27 Los bombers, 31 Karas light bombers, and numerous training and sport club planes. All were requisitioned by the Romanians and eventually used in their war against the Soviets in 1941–1944. More than a thousand aviation personnel crossed the border into Lithuania, were interned, and in 1940 were taken prisoner by the Soviets when they occupied that country.

All told, 325 Polish planes were lost, of which 116 were fighter planes. A total of 234 air personnel were lost—30 percent from bomber, 15 percent from tactical, and 15 percent from fighter squadrons—and 207 officers were taken prisoner of war. German archives at Freiburg state that 564 Luftwaffe planes were lost or so badly damaged as to be written off. A total of 189 German aircrew members were killed and 224 missing. It can probably be assumed that the Polish claims for air victories are reasonable estimates, and the balance can be credited to the very effective Polish antiaircraft defenses.

However, the richly deserved tribute to the air and ground crews has to be tempered by a critique of the high command, which proved to be anything but flexible or imaginative. There is little excuse for stationing fighter units near bomber bases (as occurred with the nearly forty planes of the Fighter Brigade) and using them for courier and shallow reconnaissance duties rather than to protect the bomber fields (in one case, this led to planes being destroyed in daylight) or the sporadic bomber operations. True, there was the prospect of getting more equipment from the west through Romania, and Smigly-Rydz hoped right up until 17 September that he could base a final stand on the Romanian "beachhead" pending the French offensive promised on the fifteenth day of the war.

Sir John Keegan commented that the Poles—lacking significant air support and mechanized forces and being attacked from two sides—held out for five weeks and fought more effectively than did the British, French, and Belgians in 1940. The Poles were defeated, but their performance certainly should not cause shame. It should be added that after the war the Polish Military Aviation was credited with being the main source of military intelligence for the Polish high command. Space does not allow a description of the heroic work of the many small units of army communication and tactical army cooperation flights that continued to work well after the major units were pulled back.

Conclusion

Polish aviation had to contend with inadequate reserves of planes and spares and frequent dislocations that consumed time and resources, thus detracting from combat operations. The latter progressively moved the base of operations east and away from the logistical centers, which were primarily around the region of Deblin (100 miles south of Warsaw on the east bank of the Vistula). At one point, special reconnaissance flights were flown to look for a gasoline cistern stranded on the railroads; then trucks were sent out to replenish gasoline. Although the Poles quite correctly and proudly assert that, with the exception of VI wing on 14 September, none of their combat units were reconnoitered or bombed by the Germans, it has to be admitted that operations from secret bases, which were only marginally able to support bomber units, impaired the effectiveness and maintenance of aircraft. German air superiority caused the destruction of lines of communication, but the final loss of bases on 17 September, when the Red Army crossed Poland's eastern border, forced the evacuation of well over a hundred Polish planes to Romania.

Critics of prewar Polish aviation policies, particularly those who believe that the aviation service was underappreciated and underfunded, ignore a basic fact. Poland's air force could operate only from land bases that were protected by ground forces. However strong the aviation arm might have been, it could not have stopped the overwhelming German military might or influenced the outcome. Paradoxically, a more powerful ground force, with a better communications system than existed in 1939 and stronger ground fortifications, had a better chance of withstanding the Germans, again assuming that a Western alliance was real. In 1944 the Germans proved in Normandy that ground forces, if anchored in place, can withstand an amazing amount of aerial punishment. In a war of movement, however, ground forces are extremely vulnerable to enemy air superiority.

The controversy over plane procurement cannot be ignored, although categorical and arbitrary opinions seem to predominate. There is no question that the PZL 37 Los bomber was the acme of Poland's fledgling air industry. It was as good as any other plane in its class. It may be regrettable that its production was canceled in 1939, but had the same monies and facilities been dedicated to the production of even the outdated (but improved) PZL 11c fighter, it might have been a different story. In its export configuration (PZL 24), this plane was sold to a number of countries, and as late as 1941, it did an excellent job for the Greeks. It was well within Poland's economic and logistical system to have tripled its fighter force in 1939 and developed a much stronger antiaircraft artillery force. Perhaps Pilsudski was right after all: what Poland needed was an aviation service that was an auxiliary of the army.

Suggestions for Further Research

Assuming that good research can answer questions and cut through polemics, the thrust of future research pertaining to Poland's Military Aviation would have to focus on the preparation for war rather than actual combat operations. But much of this research would need to be done by experts in aviation engineering and economists, since many issues pertain to these fields.

Some issues are still being argued after sixty-five years. For example, was the decision to postpone the production of PZL 50 Jastrzab fighter planes justified by the initial problems identified in test flights? The most cogent discussion seems to be by Janusz Kasprzak in *Wojskowy Przeglad Historyczny* (1981, vol. 1, pp. 323–28), but the debate continues. Also, it is well documented that General Kalkus was instructed in March 1939 to take over Polish-built planes intended for export, but this was not implemented until 1 September. Was this due to a lackadaisical approach bordering on negligence by officials in the Ministry of Military Affairs, or was it because the engines for these "export" planes were actually built in France, purchased by foreign contractors, and brought to Poland for mounting? It has also been argued that commandeering these planes would have brought criticism and possibly financial penalties and might have damaged Poland's reputation as a reliable trading partner. This too continues to be the subject of intense polemics, requiring the expert opinion of economists.

Recommended Reading

For a general understanding of the Polish experience, see Piotr Wandycz, *Lands of Partitioned Poland* (Cambridge: Harvard University Press, 1969).

For English-language histories of the Polish–Russian war of 1919–1920 and specifically the air war, see Norman Davies, *White Eagle–Red Star, The Polish–Soviet War, 1919–1920* (London: MacDonald and Co., 1972); Kenneth Malcolm Murray, *Wings over Poland: The Story of the 7th (Kosciuszko) Squadron of the Polish Air Service, 1919, 1920, 1921* (New York: D. Appleton and Co., 1932); and Janusz Cisek, *Kosciuszko, We Are Here! American Pilots of the Kosciuszko Squadron in Defense of Poland, 1919–1921* (Jefferson, N.C.: McFarland and Co., 2002).

For an account of prewar Polish foreign policy and its difficulties, see Piotr S. Wandycz, *France and Her Eastern Allies, 1919–1925* (St. Paul: Minnesota University Press, 1961), and *Twilight of French Eastern Alliance, 1926–1936* (Princeton, N.J.: Princeton University Press, 1988); Anna M. Cienciala and Titus Komarnicki, *From Versailles to Locarno* (Lawrence: University Press of Kansas, 1984); and Jozef Korbel, *Poland between East and West: Soviet and German Diplomacy towards Poland, 1918–1939* (Princeton, N.J.: Princeton University Press, 1963). For a history of the man who put his stamp on the Polish military during the Second Polish Republic (1918–1945), see Joseph Rothschild, *Pilsudski's Coup D'Etat* (New York: Columbia University Press, 1966); Waclaw Jedrzejewicz, *Pilsudski: A Life for Poland* (New York: Hippocrene, 1982); and M. K. Dziewanowski, *Joseph Pilsudski: A European Federalist, 1918–1922* (Stanford, Calif.: Hoover Institution Press, 1969).

For a brief account of the collaboration of Poland's two neighbors, see Aleksandr M. Nekrich, *Pariahs, Partners, Predators: German-Soviet Relations, 1922–1941* (New York: Columbia University Press, 1997). For a discussion of the economic issues underlying the British treasury's reluctance to advance loans to Poland, see David E. Kaiser, *Economic Diplomacy and the Origins of the Second World War: Germany, Britain, France and Eastern Europe, 1930–1939* (Princeton, N.J.: Princeton University Press, 1980), and Paul N. Hehn, *A Low Dishonest Decade: The Great Powers, Eastern Europe, and the Economic Origins of World War II, 1930–1941* (New York: Continuum International Publishing Group, 2002).

For events leading up to the Second World War, see Gerhard L. Weinberg, *The Foreign Policy of Hitler's Germany: Starting World War II, 1937–1939* (Chicago: University of Chicago Press, 1980); A. J. P. Taylor, *The Origins of the Second World War* (New York: Atheneum, 1985); Anita Prazmowska, *Britain, Poland and the Eastern Front, 1939* (Cambridge: Cambridge University Press, 1987); Simon Newman, *March 1939: The British Guarantee to Poland* (New York: Oxford University Press, 1976); *War Blue Book: Documents Concerning German–Polish Relations and the Outbreak of Hostilities between Great Britain and Ger-*

many on September 3rd, 1939 (London: HMSO, 1939); Philip V. Cannistraro, Edward D. Wynot Jr., and Theodore P. Kovaleff, *Poland and the Coming of the Second World War: The Diplomatic Papers of A. J. Drexel Biddle, Jr., United States Ambassador to Poland, 1937–1939* (Columbus: Ohio State University Press, 1976); and Anna M. Cienciala, *Poland and the Western Powers, 1938–1939* (Toronto: University of Toronto Press, 1968).

For confirmation of Polish–French staff meetings and plans and French duplicity with regard to the Poles, see General Armengaud, *Batailles Politiques et Militaires sur L'Europe: Temoignages (1932–1940)* (Paris: Editions de Myrte, 1948), 90–94. For an excellent discussion of Polish attempts to buy British planes, see Wojtek Matusiak, "The First Polish Spitfire," *Air-Britain Aeromilitaria* 29, no. 116 (2003): 169–72.

For a general history of the September campaign, see Steven Zaloga, *Poland 1939: The Birth of Blitzkrieg* (Oxford: Osprey Publishing, 2002). For the German view, see Klaus A. Maier, Horst Rohde, Bernd Stegeman, and Hans Umbreit, eds., *Germany and the Second World War,* vol. 2, *Germany's Initial Conquests in Europe* (New York: Oxford University Press, 1991), and Cajus Bekker, *The Luftwaffe War Diaries* (New York: Doubleday, 1969).

The following monograph is classic and valuable reading: Sir John Keegan, *The Battle for History: Re-fighting World War II* (New York: Vintage Books, 1996). For Poland's contribution to Enigma and Ultra, see Ronald Lewin, *Ultra Goes to War: First Account of World War II's Greatest Secret Based on Official Documents* (London: Hutchinson, 1978), and Wladyslaw Kozaczuk and Jerzy Straszak, *Enigma: How the Poles Broke the Nazi Code* (New York: Hippocrene, 2004).

Polish archives suffered irremediable losses during the war. A minuscule number of documents originating from eastern garrisons were evacuated, but most were captured by the Germans or Soviets if not self-destroyed. The Soviets returned much of their booty in the 1960s, but those archives captured by the Germans were lost in the ensuing war. In 1939 the Polish military in France attempted to re-create from memory the relevant data surrounding events prior to and during the September campaign. This led to the most comprehensive Polish-language report: Olgierd Tuskiewicz (chief editor), *Lotnictwo w Kampanii Wrzesniowej* (unpublished Polish Air Force Historical Section report; copy in author's library). All this material is in the Archives of the Polish Institute and General Sikorski Museum in London and is open to bona fide researchers. As a result, a number of books have been published; see Adam Kurowski, *Lotnictwo Polskie w 1939 r* (Warsaw: Ministerstwo Obrony Narodowej [MON], 1962), and Ryszard Bartel, Jan Chojnacki, Tadeusz Krolikiewicz, and Adam Kurowski, *Z Historii Polskiego Lotnictwa Wojskowego, 1918–1939* (Warsaw: MON, 1987). A number of archival documents that survived were published as *Wojna Obronna Polski, 1939. Wybor Zrodel* (Warsaw: MON, 1968). For a good reference to Polish aircraft construction, see Jerzy B. Cynk, *Polish Aircraft, 1893–1939* (London: Putnam, 1971).

The following memoirs were pertinent to this historical essay: Jozef Zajac, *Dwie Wojny, 1914–1939* (London: Veritas, 1964); Sir John Slessor, *Central Blue* (London: Cassell and Co., 1956); Roderick McLeod and Denis Kelly, eds., *Time Unguarded: The Ironside Diaries, 1937–1940* (New York: David McKay, 1963); and Alfred B. Peszke, "The Bomber Brigade of the Polish Air Force in September, 1939," *Polish Review* 13, no. 4 (1968): 8–100.

For one of the few English-language books on prewar Polish aviation, illustrated with rare photographs, see Bartlomiej Belcarz and Robert Peczkowski, *White Eagles: The Aircraft, Men and Operations of the Polish Air Force, 1918–1939* (Ottringham, England: Hikoki Publications, 2001).

For a comprehensive listing of all published material on the Polish Military Aviation and air force, see Michael Alfred Peszke, "Review of English Language Historiography of the Polish Air Force," *Air Power History* 50, no. 1 (2003): 46–49.

CHAPTER TWO

L'Armée de l'Air, 1933–1940

Drifting toward Defeat

Anthony Christopher Cain

Between 10 May and 25 June 1940, Germany overran France. The Armée de l'Air, the French air force, fought ineffectively against Germany's Luftwaffe. Pierre Cot, two-time air minister under the Third Republic, declared, "The easy conquest of France by the Wehrmacht in May and June 1940, was largely due to the perfect collaboration of armored divisions and aerial forces. . . . Such excellent co-ordination of effort and movement cannot be improvised on the battlefield; it is the result of minute preparation, long training of soldiers, and especially of the adaptation of the military organization and machinery." One of the most neglected areas of twentieth-century historiography centers on how a powerful nation like France allowed its air service to wither away to the point of such ineffectiveness. In other words, how could Germany manage to orchestrate such "minute preparation" while France failed to do so?

Historians have judged the French military—and the Armée de l'Air in particular—as not being up to the task of defending France in any strategic, operational, or tactical dimension. Despite the burdens of a punitive peace, economic depression, and a materiel-starved military, a small cadre of German military professionals built a military machine that burst onto the international scene in 1939–1940 and succeeded in outthinking, outplanning, outbuilding, outtraining, and, ultimately, outfighting their French rivals. In other words, France squandered twenty years of preparation and investment, while Germany capitalized on every opportunity to defeat its longtime adversary. There is a degree of justification for this judgment. French politicians, strategists, and military professionals regarded Germany as the most pressing security threat almost from the outset of the post–World War I era. With Hitler's rise to power, that threat stood out in sharp relief. Even if one considers that influential members of the French government hoped that international disarmament initiatives would avert war, considering the stakes, operational and tactical plans should have evolved toward greater flexibility.

The Armée de l'Air received a disproportionate share of the blame for the 1940 defeat compared with the army and navy. There was no shortage of accusers in the aftermath of defeat. Anatole de Monzie declared, "We lost the war because of aviation. We lost for other well-defined reasons, but its tragic obviousness haunts the spirits of Frenchmen. We do not hesitate to speak of aerial treason." Historian Marc Bloch established an enduring image of French airpower's ineffectiveness in his famous postmortem of the war, which described the shock and hopelessness of the French infantrymen who experienced unrelenting German aerial attacks: "The fact is that this dropping of bombs from the sky has a unique power of spreading terror. They are dropped from a great height, and seem, though quite erroneously, to be falling straight on top of one's head. . . . Exposed to this unleashing of destruction, the soldier cowers as under some cataclysm of nature, and is tempted to feel that he is utterly defenceless—though in reality, if one dives into a ditch or even throws oneself flat on the ground, one is pretty safe from the bursts."

Marshal Philippe Pétain's Vichy government launched a trial to identify and condemn those who were primarily responsible for the disaster. The court indicted such leaders as Prime Ministers Léon Blum and Édouard Daladier and Air Ministers Guy La Chambre and Pierre Cot for, among other things, their roles in failing to prepare the Armée de l'Air to stop German aggression.

One author describes the air service as a scapegoat for the defeat—an institution on which politicians and army leaders could conveniently heap accusations to divert attention from their culpability in the disaster. The loss of air superiority and the air force's inability to regroup left French ground forces exposed to a three-dimensional combined-arms assault that paralyzed operational and tactical commanders. The air force's failure to create an organization, doctrine, and tactics in the years leading up to the war that would have allowed it to effectively counter the Luftwaffe was a fatal mistake that accelerated the French army's collapse after the Germans crossed the Meuse River on 13 May 1940. One air reconnaissance commander observed that the French seemed to have matters under control until the Germans forded the Meuse at Semoy. "On the morning of the 14th, Captain Vachette [a reconnaissance group pilot] reported that the 55th Infantry Division had abandoned its lines." From that point on, French forces lost the ability to present an effective defense against the Germans' rapid combined-arms attack, and the cascading collapse accelerated as lower-level units lost touch with higher command echelons.

If one agrees that the Armée de l'Air was primarily responsible for the defeat, one must accept that politicians and military leaders had created effective mobilization, logistical, and war-fighting systems designed to support a winning strategy. This was certainly not the case, and air force leaders should receive harsh criticism for their failures in these areas. To justify the degree of blame customarily leveled against the air force and its leaders, however, one would have to accept that France's defensive strategy, predicated on meeting the German assault in the Low Countries while holding the formidable Maginot Line, would have achieved meaningful political and military goals and preserved French sovereignty. For the strategy to work at all required a rather sophisticated degree of coordination between land and air components that was absent. If one assumes that the strategy was valid, ground commanders should have used the protection afforded by airpower to detect the main thrust of the German attack and then shift their forces to stall the Wehrmacht's advance. However, the German high command did not settle on a strategy until well after France and Great Britain declared war in September 1939. Thus, the French could not have foreseen the exact outlines of the German thrust through the Ardennes before it occurred. Further, to validate the scapegoat thesis, one would have to discover evidence indicating what the French intended to do—using airpower as a primary strategic and operational element—to conclude the war after blunting the initial German advance. All these assumptions are suspect to some degree, and failure in any area probably would have produced a defeat similar to the one that occurred in the late spring of 1940. In other words, achieving a stalemate is very different from achieving a victory; the French high command failed to look beyond the initial defensive battle to devise a war-winning strategy. Without such a strategy, a more effective use of airpower would not necessarily have changed the outcome to France's advantage.

A critical investigation into the roots of the Armée de l'Air's role in the 1940 disaster involves examining the strategic, organizational, technological, and operational factors that produced the unique performance characteristics that led to the defeat. Such an inquiry should not seek to exonerate the Armée de l'Air and its leaders; it should, however, reveal strategic, operational, and tactical choices that caused the French air force to fail to execute its basic mission and thereby justify the expectations of the French people. More important, a rigorous analysis of the Armée de l'Air's performance in 1940 should lead scholars to ask how, equipped with the information available to the leaders of the day, the air service could have performed better.

Strategic Roots of Defeat

Analyses of the strategic roots of the 1940 disaster generally contrast the French vision of modern war with that of their German adversaries. Following this line of reasoning, authors usually point to the French adoption of fixed defensive positions and highly centralized command-and-control mechanisms, whereas the Germans opted for mobile combined-arms thrusts that capitalized on decentralized small-unit initiative and leadership to isolate inflexible French tactical and operational units. Such accounts emphasize the investment in the sophisticated fortifications of the Maginot Line and the doctrine of methodical battle as key weaknesses in the French strategic posture. If French army leaders learned one lesson from the battles of World War I, they learned that defensive preparations backed up by concentrated firepower caused the enemy to expend his forces while they conserved theirs. Even the most prominent airpower theorists of the interwar years—Italian Giulio Douhet, British Sir Hugh Trenchard, and American William "Billy" Mitchell—expected land warfare to stagnate into attritional trench warfare owing to the lethal effects of firepower. Analyses of how land warfare doctrine evolved advance our understanding of why French land forces fought the way they did, but they do not provide sufficient insight into how the Armée de l'Air performed, nor do they help arrive at judgments regarding air service culpability in the ultimate outcome.

French strategy was defensive and reactive for most of the 1930s. French leaders attempted to surround their adversary with a series of alliances that would present multiple threats to German strategists and diplomats after it became obvious that the proscriptions of the Treaty of Versailles would be inadequate to constrain German power. The Armée de l'Air figured prominently in these strategic alliance-building initiatives. France relied on Great Britain to deploy forces across the English Channel, as it had in 1914, but by 1939, the consensus that had carried England into the First World War was somewhat weaker. Despite French perceptions that their allies from across the Channel did not do enough, the British responded bravely and faithfully to the French call for assistance, but German technology, strategy, and tactics had changed the context—there would be no "miracle on the Marne" for France in 1940. As far as British airpower was concerned, the glory days of the Royal Air Force (RAF) were in the distant future. Though Britain sacrificed light and medium bombers alongside its French aviation brethren over the skies of northern France and the Low Countries, the British government was unwilling to squander valuable fighter planes and irreplaceable pilots that would later be sorely needed and tested as they defended Britain against the Luftwaffe.

In eastern Europe, French strategy drifted from one failed alliance to another. Leaders on the Left, including Cot, believed that the massive military potential represented by the Soviet Union, combined with a resolute Franco-British alliance in the west, could present Germany with the dilemma of an unwinnable two-front war and thus put the brake on Hitler's territorial ambitions. The Soviets, however, proved to be difficult to entice into a stable alliance. They consistently demanded more than French diplomats were willing, or able, to give. From the Soviet perspective, the French appeared unreliable. When faced with a choice between provoking Germany by signing an alliance with France and signing an agreement with an untrustworthy Germany, Stalin opted to add to Soviet power, if only temporarily, by making a deal with the known evil rather than risking all on questionable French promises of assistance.

French diplomats also courted the smaller, more vulnerable Eastern European countries—particularly Poland and Czechoslovakia—with the same strategic goals they hoped to achieve through an alliance with the Soviet Union. In essence, the French reliance on airpower as an instrument of foreign policy was consistent with 1930s movements that favored collective security and disarmament. By placing a ring of threats around Germany's eastern borders, French leaders hoped to divert attention from their own weaknesses and thereby preserve the peace. They offered airpower as a guarantee to their erstwhile eastern partners—French air fleets would shuttle between their home bases and fields in Poland and Czechoslovakia while raining destruction on German targets en route. After rearming and refueling at allied airfields, they would return to France after again dropping their bomb loads on lucrative targets in Hitler's Reich.

Cot pushed the collective security argument further and suggested that nations disband their individual air fleets in favor of an international air force—under French command, of course—that could be deployed rapidly to bombard aggressor nations, thus guaranteeing the peace without allowing nationalist factions to push their states toward another global war. He wrote:

> [The] interallied air force, formed of Divisions or Squadrons from different nations, would have three great missions. For one part, to reinforce the defenses of essential political centers whose destruction could influence the conduct of the war; thus the integrity of Paris or London is important not only to France or

> England, but to the entire coalition. Second, to prevent the rapid encirclement of the Czechoslovakian bastion and to permit the arrival of Russian troops to confront the German forces. . . . Finally, and above all, the Interallied Air Force should carry out massive aerial reprisal and bombardment missions.

Of course, the emphasis on protecting Paris rather than Prague or Warsaw communicated an ulterior motive behind Cot's proposal. The problems with such a strategy became obvious in the late 1930s as powerful French air fleets proved nonexistent and the German air force emerged as a potent combatant and a contender for air superiority in its own right.

For all its theoretical potential, French airpower could not rise to the heights that diplomats demanded of it to deter German aggression. The fatal flaw in French attempts to constrain Germany using airpower as a deterrent centered on the lack of materiel and moral commitment behind the French promises. Deterrence strategies surrendered the initiative to Germany; they worked only as long as German leaders believed that French leaders were willing to go to war to achieve their security objectives and as long as German leaders were unwilling to gamble that France would not call their bluff.

The alliances represented a way to provide cheap security for France; they created the illusion of safety while ensuring that, if war came, French blood would not be the first spilled. Additionally, the offer of aerial bombardment as France's contribution to collective security appeared to be a significant one that carried great psychological weight, but French diplomats knew that their country had failed to make the necessary investments to make good on their promises. French security guarantees proved hollow for alliance partners. Ultimately, France and Great Britain were unable and unwilling to come to the aid of their Czech ally and unable to intervene in time to rescue Poland from the German onslaught. General Joseph Vuillemin, chief of the air force General Staff, communicated the bankruptcy of the France's reliance on airpower as a strategic deterrent when he remarked in 1938 that, in the event of war, the Luftwaffe would destroy his service in two weeks. Tragically, events revealed that France could not save itself, much less protect its allies from German aggression.

Organizational Roots of Defeat

French airpower strategists offered no better options when they attempted to solve the problem of an aggressive Germany from an organizational perspective. Although not readily apparent at the time, the 1933 decree that separated the Armée de l'Air from the other services contained the roots of defeat. In an organizational decision that modern U.S. military strategists would recognize as promoting "jointness," the Chamber of Deputies in 1934 passed a law that directed the air service to develop abilities to "participate in aerial operations, in combined operations with the Army and the Navy, and in territorial defense." Problems with this approach became apparent when airmen attempted to articulate their vision of how the air force's unique capabilities could contribute to the national defense. Army and navy leaders resisted aviators' attempts to describe how airpower could transform traditional battlefield missions. In retrospect, it is clear that this law gave the army and the navy leverage over the new service to ensure that little changed with respect to strategic and operational priorities.

Pascal Vennesson argues that the power imbalance among the services forced the air force into a subordinate role in strategic and operational decision-making councils. According to Vennesson, this kept the air service focused on tactical issues that mattered to the army and the navy rather than on developing doctrine, tactics, techniques, and procedures that met the needs of operational air warfare. The strategic decision to emphasize the air force's supporting role for land and sea forces without establishing a clear priority among the three statutory missions kept the service in a perpetual reactive posture. Furthermore, because the air service was relegated to a supporting role for the army, airmen were forced to adapt their visions of the pace, scope, and nature of air warfare to better suit the methodical battle favored by the army. This would become evident as the air service began to build and train the operational force it would use to fight the coming war.

A word of caution is in order here. One could criticize the French for choosing ineffective organizational models—for choosing to limit the Armée de l'Air's independence by directing it to support surface forces. At the time, however, the fear that airmen would use their newfound autonomy to pursue independent missions at the expense of the other services' support requirements was reasonable. Airpower theories varied regarding the appropriate targets (materiel, morale, or a combination of both), but most emphasized using airpower as an independent force launched against targets beyond the battlefield. Although prevalent airpower theories of the day held that strikes against urban areas represented the optimal use of airpower, in the 1930s, few nations committed the resources required to build potent strategic strike forces.

French legislators were understandably reluctant to allow the air force to develop and implement plans for deep strikes into Germany while leaving vital Gallic social, political, and economic targets at the mercy of enemy bombardment fleets. Additionally, in broad terms, the French organizational philosophy that emphasized air support for ground operations resembled the one employed by the Germans during their years of conquest. The Luftwaffe experimented briefly with the idea of long-range strategic bombing, but the imperative of a short war, materiel scarcity, and pressure from ground commanders forced German airmen to place theories of victory through independent air operations on the shelf. When the tide turned against Germany, Hitler's airmen found themselves in much the same position that the French encountered in 1940. The Luftwaffe proved no better at creating a more effective, balanced air force under the constant pressure of combat in 1943–1945 than the French had in 1940.

The organizational problems rooted in the constraints of the law that established the air service as an independent entity manifested themselves in ill-defined roles and missions. Satisfying army and navy demands for aerial resources while addressing the potential for independent air operations became the most consistent problem as airmen attempted to establish an institutional identity and a legitimate role for their service. The air service's ability to provide persistent reconnaissance and surveillance over operational and tactical areas sparked a constant battle for scarce aerial resources as operating concepts evolved during the 1930s. Army and navy leaders assumed that they had priority for air support, as well as the institutional power to ensure that airmen honored their demands. The Armée de l'Air became merely a junior competitor in the power struggles to prioritize roles and missions for aerial resources.

For much of the 1930s the Armée de l'Air found itself buffeted between organizational models that alternately concentrated aerial resources and dispersed air units to support the army. Cot argued for concentrating air forces under geographic commands. Under his influence, the service deployed five air armies, with two focused on guarding the northern and northeastern areas that would be most threatened in the event of a German attack. A third army shielded the southeast and the Mediterranean area, a fourth provided reserves for the Métropole (metropolitan France), and a fifth aerial army guarded the North African colonies in Algeria and Morocco. Under this organizational scheme, air commanders could apportion the capabilities at their disposal to all three statutory missions. Airmen could find themselves tasked for pursuit (air superiority), attack (ground support), bombardment (interdiction and strategic attack), reconnaissance, surveillance, or liaison (command, control, and coordination) missions. Critics of the geographic command organization argued that the air staff could ignore or assign lower priorities to missions that emphasized support for land forces.

The dispersed organizational model favored the army's concerns that German airpower would dominate the battlefield. Under this model, operational and tactical army commanders controlled priorities for employing aerial missions. In effect, each army unit wanted its own air element to support operations in its immediate battle area. This resulted in an emphasis on reconnaissance, surveillance, artillery spotting, and liaison missions at the expense of a wider range of airpower possibilities.

A third organizational complication manifested itself in the form of territorial air defense missions. The Défense aérienne du territoire (DAT) organization represented another dispersal of air resources while providing nominal security for important urban areas. Encompassing pursuit, ground, and civil defense missions, national mobilization plans called for the DAT to revert to Armée de l'Air command during combat operations. In reality, the peacetime organization of DAT forces never meshed well with air-service war planning or operations. Reservists constituted the majority of DAT personnel, and they never attained the proficiency required to detect and respond to potential enemy attacks and pass the information to higher echelons; in addition, their equipment never measured up to frontline standards. Therefore, what appeared to be, on paper, a rational system intended to provide a layered air defense for vital towns and industries while capitalizing on the cost-effectiveness of having a ready reserve system never measured up to the requirements of its assigned mission.

Cot attempted to move the DAT mission away from the air service during his first term at the Air Ministry (1933–1934) by offering to give the army light pursuit aviation units, which had traditionally performed the DAT mission. The move was supposed to appease army demands for air support, make the remaining air resources available for wider operational missions, and eliminate aging and obsolete equipment from the air service's inventory. The strategy backfired. Instead of easing army demands for air support missions, the move gave the land component a precedent for increasing its demand for control over air resources; in the minds of ground commanders, it confirmed that they should own and operate their inherent aviation assets.

The dispersed organization employed by the Armée de l'Air in 1940 eased army concerns regarding control of aviation assets only temporarily. Even before the attack began, French commanders realized that they could not assemble enough aircraft to meet the German fighter, bomber, and close air support attacks at critical points. The experience of the V Armée proved typical. The Armée de l'Air provided approximately fifteen aircraft to guard a 200-kilometer front. The alerting network was so inefficient that by the time observers detected approaching German airplanes, formatted the proper messages, forwarded these to the appropriate headquarters elements, and scrambled the nearest French pursuit units, the

Germans would have come and gone. According to one report, "The Groups are too far [from the areas they support] to be able to act in time against a massive aerial attack which could be very short and will never lack the element of surprise." In other words, the Germans would benefit from their ability to mass their formations over critical target areas, while the French dispersed organizational scheme would prevent them from massing defensive resources over the same areas.

Between 1934 and 1940, air service leaders attempted to balance the requirements imposed by the 1934 law against their views of how to employ airpower most effectively. At nearly every initiative, they found that domestic and interservice politics stalled their efforts. By February 1940, the organizational battle had come full circle; Air Minister Guy La Chambre and Chief of Staff Joseph Vuillemin agreed to dismantle the regional air command structure in order to push air support to army units near the front. This effectively placed the air service in an organizational relationship similar to the one that had prevailed during the First World War—army commanders used airpower as another combat arms branch, and aviation became an extension of the cavalry and artillery. Tactical and operational army commanders tasked airmen to perform missions, but they lacked an effective centralized command structure to orchestrate the aerial effort across the entire theater. One officer bitterly related this shortcoming: "In approximately one month, the air forces of the X Corps d'Armée changed command five times. The shifting of the headquarters required because of the prevailing conditions proved prejudicial to providing effective support to the Corps d'Armée . . . the result was an almost complete lack of understanding on the part of land commanders of even the most modest capabilities of the Air Force including reconnaissance, pursuit, or bombardment resulting in demands that were impossible to satisfy." In doctrinal terms, there was no unity of command focused on directing the full range of airpower capabilities to achieve operational and strategic objectives. Even worse, when the Germans broke through the French positions, the absence of a centralized planning and coordination capability for theater air resources meant that army leaders focused on ground component issues first, failing to use the air resources at their disposal. This meant that airmen had to fend for themselves for even the most basic operational and tactical necessities in the ensuing chaos of operational collapse.

Technological Roots of Defeat

Air forces are like navies, in that they often rely on and identify with technology more than their land-force counterparts do. Effectiveness in the air can boil down to which combatant quickly achieves the best combination of technological innovation, training systems, doctrine, and organization; in this sense, effectiveness is often a matter of comparative rather than absolute advantage. In the 1930s nearly every major power witnessed an aeronautical technical revolution that continued throughout the war. By the end of the decade, most air forces had abandoned biplanes in favor of monoplane designs; engine and fuel technology had improved to provide dramatically higher output. Wing designs afforded a better lift and drag ratio that resulted in higher operating ceilings and heavier payloads. Technological achievements, however, required risky investments to keep a nation's air force at the peak of modernization. In 1930s Europe, fiscally conservative lawmakers were reluctant to risk public money on costly, unproven aviation technologies that might turn out to be only interim steps in revolutionary design developments. France drifted through the latter half of the 1920s following an air production strategy known as the *politique des prototypes,* which offered official encouragement to aircraft designers but provided few substantial orders for combat airplanes. By 1933, the once vaunted French air capability had atrophied.

Hitler's rise to power provided French leaders with the incentive to abandon the prototype policy in favor of a true operational air force. But although the German threat supplied focus and purpose, it did not furnish the stimulus for France to create a comprehensive production strategy to revive its moribund air capability. The global economic crisis was finally being felt in France after some deceptive grace years. Aircraft manufacturers operated artisanal industries that were not prepared for the demands of a sustained industrial mobilization, owing in part to the abysmal salaries they paid their workers—in 1938 skilled aircraft workers earned an average of 7 francs per hour (21 cents per hour in 1938 dollars). The government had not provided enough incentive for designers to innovate or even to match the pace of state-of-the-art developments. Hitler's bellicose rhetoric and aggressive industrial programs created an oppressive atmosphere laden with the anticipation of impending war. Finally, the law that had created the air service constrained procurement policies. The Air Ministry could not purchase bombers at the expense of reconnaissance or pursuit aircraft without incurring the wrath of the other services.

Pierre Cot and his chief of staff, General Victor Denain, made several decisions intended to solve these complex technological problems, and they would haunt the Armée de l'Air for the remainder of the 1930s. First, they chose to build a series of multipurpose aircraft that would put the air service on a par with its competitors. Second, they attempted to become more directive toward French aircraft manufacturers to acquire greater control over the scope and pace of aircraft design and production. Finally, they threatened to nationalize the French aviation industry to provide government control over production and innovation. Cot attempted to carry out the nationalization threat during his second term as air minister under the Popular Front government (1936–1938), with predictably disastrous results.

The modernization program began as an attempt to provide the most cost-effective technology while also convincing the other services that the Armée de l'Air would continue to meet their air support needs. The resulting series of airplanes, called the Bombardement, Combat, Reconnaissance (BCR), reflected the compromises that led to their design. They were bulky, slow, and ill suited for any of the missions from which they took their name. By 1936, French analysts realized that the day of the multipurpose aircraft had passed. One author wrote: "Once a mass formation of heavy aircraft has been broken up by combined anti-aircraft and pursuit attack, it would be very difficult for the cumbersome ships to reform rapidly into a compact unit. . . . The single-seat fighter has made such progress in speed and armament that it has entirely revolutionized methods of aerial combat just at the time when the Douhet type of heavy cruiser might have been expected to bear fruit."

The master procurement strategy designed to modernize the Armée de l'Air with the BCR series envisioned placing 1,010 airplanes in service by 1936—in time to equip the air service with the world's most modern aircraft to counter an anticipated German attack. From the beginning, French industry failed to meet the goals of the production plan. The economic crisis had poisoned relations between management and factory workers—relations that had never been healthy in France's politically charged labor sector. And the years of small-scale production meant that conditions in the aviation industry were better suited to building World War I–style aircraft rather than mass-producing modern, all-metal monoplane designs. Moreover, French designers tended to build airplanes without considering the various missions the planes would perform; thus, the basic airframe may have met the requirements for a bombardment mission, but the manufacturer often failed to provide bombsights or even basic flight instruments. The BCR aircraft were probably better suited for reconnaissance missions than for bombardment or aerial combat. Despite these obvious shortcomings, the air service would not begin to acquire specialized aircraft—and even then, in numbers too small to influence the Armée de l'Air's combat potential—until the eve of war with Germany.

Controversies over costs and missions precluded midcourse corrections in the BCR program. The inherent instability in the Third Republic government and France's pursuit of international disarmament combined to constrain aircraft production further. The dilemma was partly one of the air staff's own creation. On the one hand, Cot and Denain had campaigned so aggressively for the BCR program that they could not reverse course and admit that the theory of a multipurpose battle plane was technologically bankrupt. On the other hand, the Armée de l'Air leaders could not win another battle for a major rearmament program in the face of resistance from those who favored disarmament, with whom Cot identified intellectually. France could not speak out publicly for disarmament while it privately launched programs to produce air fleets capable of projecting offensive power across Europe. Thus, the air service found itself saddled with the technological and political baggage of the BCR program until Guy La Chambre became air minister in 1938. By that time, however, French designers and producers could not overcome the misconceptions of the BCR era.

The Amiot 143 was typical of the early BCR program. Its high-wing, twin-engine, monoplane design gave it performance characteristics that compared favorably with the American Martin B-10 and later the Douglas B-18 bomber designs of the early 1930s. Two Hispano-Suiza engines delivered 880 horsepower that resulted in a cruising speed of 190 miles per hour. The Amiot, though bulky, did not have enough armor or speed to allow it to provide reconnaissance or attack support for ground forces. Its 372-mile combat radius meant that it was not the deep-striking weapon that French diplomats had promised their Czech and Polish allies. When Armée de l'Air leaders threw the obsolete BCR stereotype into the fight near the Meuse in 1940, more than half the airplanes and their crews never flew again.

The persistence of the BCR legacy became evident when products from other aircraft manufacturers began to enter the French air service. Between 1934 and 1938, Farman, Bloch, and Potez delivered aircraft that conformed to the basic design and performance characteristics of the Amiot 143. Average speeds of 200 to 210 miles per hour and combat radii between 330 and 600 miles were the norm. This placed French aviators at a 100 mile per hour disadvantage compared with their German adversaries equipped with Messerschmitt Bf 109s.

When the Air Ministry approved plans for airplanes specifically designed for pursuit missions—the Morane-Saulnier MS-406 and the Dewoitine D-520—French industry could not shift its production emphasis in time to tip the scales in their favor. At the beginning of the campaign in May 1940, the MS-406 was the Armée de l'Air's most numerous modern fighter, with approximately 500 deployed to combat units as of October 1939 and an additional 500 on order. The Bf 109, with its nearly 60 mile per hour speed advantage, wiped out nearly half the MS-406 inventory during the campaign. The French responded by attempting to rush the D-520 into service. They began the campaign with 36 D-520s in service and an additional 194 on order. The pressures of combat and the national disaster complicated the usual problems that accompany the deployment of a complicated major weapons system. By the end of the Battle of France, 85 of the graceful D-520s had fallen in combat or in accidents.

The constant battles among aircraft manufacturers, labor, and the government were no secret. Throughout the 1930s, Armée de l'Air leaders tried unsuccessfully to entice manufacturers to relocate their facilities away from the vulnerable Île-de-France area around Paris to locations in the south and southwest that offered better security. Lawrence Bell, of the U.S. Bell Aircraft firm, toured Amiot, Morane-Saulnier, and Potez factories in the summer of 1938 and filed a report of his observations with the U.S. Army Air Corps. According to Bell, conditions at Amiot and Morane-Saulnier were similar. He described the Amiot plant as "very poor, low headroom, inadequate light, untidy and most of the equipment is obsolete, no really modern production equipment or methods. . . . The employees did not look industrious, discipline was poor and morale appeared to be low." Such a negative critique of any manufacturer would have been cause for concern, but its applicability to the Morane-Saulnier firm, responsible for producing the MS-406, embarrassed the Air Ministry. Bell reported, "If that part of the industry that was not inspected is no more active or efficient than the three plants visited, the state of the military aircraft industry in France is pretty low."

Air Ministry leaders took Bell to a recently nationalized Potez plant responsible for producing the Potez 63 light bomber. Officials told Bell that the Potez facility was "the show plant of France." Although the production floor at Potez seemed to be cleaner, more organized, and better maintained than at the Amiot and Morane-Saulnier factories, Bell noted that production processes did not measure up to the demands of rapid industrial output. "Of the first 100 planes of this type [Potez 63] being completed on the assembly floor, the majority are held up waiting engines . . . the engine manufacturer's [Hispano-Suiza] export business holds up domestic production." From a strategic perspective, the Potez plant's location in northern France made it an inviting target for German attack.

As war approached in the late 1930s, French military aviation industry leaders did not have the resources or the momentum to close the gap between their country's needs and its capabilities. Between June 1937 and January 1938, the French aviation industry delivered only 71 combat-ready aircraft, while German firms produced 4,342, Britain produced 2,335, and the United States produced 293. These numbers do not tell the entire story, however; the types of airplanes manufactured by Germany and France were just as important to their ability to fight a war. U.S. Air Corps Intelligence Service analysts estimated that by June 1939, Germany had 3,203 pursuit aircraft, of which 2,978 were capable of speeds of 320 miles per hour or greater. The French had only 1,175 pursuit aircraft, with only 11 capable of competing with the German high-performance fighters. The comparison did not favor the French in any category. Germany fielded 552 attack airplanes; the French inventory contained 378. Germany had 5,375 medium and heavy bombers; the French had only 997. Germany equipped its forces with 1,443 observation aircraft, while France fielded only 424. When the two adversaries faced each other, Germany had a better than three-to-one advantage in total airplane numbers (10,871 to 3,024), with the most critical difference being in high-speed pursuit and bombardment categories.

Operational Roots of Defeat

The Armée de l'Air's operational theory of air warfare represented an attempt to reconcile differences among competing strategic, organizational, and technological factors that shaped the context of the 1930s. French airmen were familiar with prevailing airpower theories of the day, and they drew on them—particularly the ideas of Douhet—to create their own operational formula for victory in the form of a published doctrine. Although air service leaders articulated a sophisticated vision of how to employ airpower as both an independent and a supporting force, they failed to forge a fighting force that could translate the vision into operational reality.

In large part, Armée de l'Air leaders assumed that land warfare would conform to the patterns and pace of World War I—combatants would be incapable of achieving mobility or advantage on the battlefield owing to the effects of fortification and firepower. This meant that airpower would set the conditions for victory by attacking the enemy's operational forces, his second-echelon reinforcements and supplies, and his homeland. *Lutte aérienne* was the doctrinal term French airmen used to characterize how airpower had changed modern warfare. It communicated a vision of a seamless battle space in which offensive and defensive campaigns occurred simultaneously. This conception allowed airmen to reconcile their legal requirements to support surface forces with theories that emphasized airpower's inherently offensive nature. Although the published doctrine represented an attempt to articulate a comprehensive view of the nature of air warfare, it emphasized tactical aspects at the expense of operational or strategic capabilities.

The operational force structure that resulted from the *lutte aérienne* doctrine appeared to be a rational, balanced force with the flexibility to shift levels of emphasis from offense to defense as the situation demanded. When combined with the technological philosophy behind the BCR program, this doctrinal structure should have produced one of the most sophisticated aviation capabilities of the interwar period. As discussed earlier, the technological failures of the BCR program negated one aspect of the doctrine. Even if French aircraft firms had produced the requisite numbers of effective aircraft, the Armée de l'Air failed to translate the concepts contained in *lutte aérienne* into operational reality. Almost from the outset, schools, training programs, and personnel structures proved inadequate for the needs of a modern air force. Doctrinal manuals began by emphasizing that morale and attitude were essential to victory. The authors of the *Règlement de Manoeuvre de l'Aviation* wrote: "The will to conquer is the essential factor for success . . . to inspire this will and to teach the means of victory is the object of instruction."

Doctrine and educational programs aimed to equip individuals and the squadrons they served with a "veritable passion for flight" and an understanding of the "evolution of ideas concerning the construction of airframes and motors in France and abroad." In the fiscally constrained environment of the 1930s, however, doctrinal instructional programs stressed ground courses as the foundation for subsequent flight instruction. The air staff divided the annual training calendar into blocks that allotted considerable time to studying land and naval warfare concepts so that airmen would be thoroughly familiar with their duties when called on to support the senior services. When ground preparation and instruction integrated aviation concepts, it occurred in the context of "the employment of the air force, to specify the situation of land or naval forces, and the orders given to the troops in the echelons studied."

Because of the prominence of multipurpose aircraft in the Armée de l'Air, a great deal of instruction focused on crew and formation training. Individual proficiency training progressed from basic qualification through various stages leading to fully mission-qualified and instructor-qualification status. Doctrine prescribed stability among the crew force—maintaining integral crews and emphasizing the role of formation leaders—as the ideal way to manage individual and crew qualification. In this way, a common aviation culture emerged; "the crew members know each other as perfectly as possible [they are] accustomed to serving together under the orders of the same aircraft commander, possessing one soul, aware of and proud of their own value, taking total care to conserve the equipment with which they are entrusted." In actuality, the emphasis on aircrew discipline and stability created a culture that lacked initiative. Thus, Armée de l'Air leaders envisioned creating a service capable of adjusting to fluid combat conditions in predictable ways under the common structure of crew, squadron, element, and group organizations. Theoretically, as the inevitable losses of combat occurred, aviators would retain combat effectiveness because of their familiarity with their institution's established doctrine.

Although the concept of a seamless air battle encapsulated in the structure of *lutte aérienne* contained both offensive and defensive aspects, it devoted more emphasis to aviation's supporting role for surface forces. In this construct, airpower concentrated on detecting and intercepting enemy aviation attacks aimed at French land forces. This meant that in its primary role, French airpower assumed a defensive posture that surrendered the initiative to enemy air forces. In other words, doctrine assigned aviation a role that stressed using its inherent range, speed, and flexibility to "adapt its actions to the enemy," rather than seizing the initiative to complicate enemy operations.

One could view this prescription as a sophisticated attempt to solve the problems of firepower and maneuver that theorists expected to occur when armies clashed. Airpower, with its ability to envelop the enemy vertically, could become the only possible maneuver element available to national leaders. As air doctrine noted: "Only airpower is equipped to cross the impenetrable fronts on terrestrial reconnaissance missions, quickly collecting necessary information at long range. It is upon airpower that the security of the armies rest. . . . The Armée de l'Air should be consumed with the importance of the mission which in this way is reserved for aviation, and in which the best execution is indispensable to the success of the land battle." The problems of operating over and across the anticipated battle areas became insurmountable because, as discussed earlier, French industry proved unwilling or unable to produce airplanes that would carry the fight to the enemy homeland. The inadequacy of French pursuit planes in the face of high-speed, highly maneuverable German fighter aircraft further compounded the problem; the French pursuit planes needed speed and loiter time (both of which they lacked) to react to penetrating German formations. This doomed the French army to fight under constant pressure from German aviation while the French air force appeared to surrender the skies to the enemy.

By 1939, France had the second largest air force among the major powers, with 213,654 personnel; however, if the reserve component was subtracted from the total, the Armée de l'Air ranked only sixth. But the number of active pilots available for air service is more telling. By November 1939, the Armée de l'Air could count only 2,790 officer and

2,989 warrant officer pilots among its ranks. Since 1933, when the air service gained its independence from the army and navy, the Armée de l'Air's leaders had failed to create a training system capable of preparing aviators to man cockpits and mechanics to service the machines. Consequently, the Armée de l'Air's inability to gain air superiority during the German assault sealed the fate of its crews and its homeland.

Faris R. Kirkland has noted that the high command deliberately made a miscalculation in 1936 when deciding on personnel policy, in contrast to the RAF's and U.S. Navy's 1934 decisions to expand both air and ground personnel. As a result, many French aircraft could not be manned and maintained and so were stored on airfields all over France. According to Vuillemin, the French aircraft industry produced 358 planes in January 1940, but the Armée de l'Air accepted only 198 because the rest were unusable; they were not properly equipped, and the necessary personnel did not exist.

Alternatives to Defeat

Exploring counterfactual history is always a risky business. French soldiers and airmen did not expect to lose the war against Germany in 1940; thus, historians must view their preparations during the 1930s with an appreciation of the context that shaped their decisions. These decisions had a rational basis that passed the reasonability test among government leaders and military professionals alike. Nevertheless, at several points in the 1930s, French leaders could have chosen other options. The most important areas under direct Armée de l'Air control were technology, organization, and operations.

Technologically, the decision to procure the BCR aircraft series to modernize the Armée de l'Air by 1936 proved fatal. Although the decision soothed interservice political concerns, it provided modern technology only briefly. Production problems and an ongoing technological revolution saddled the air service with useless materiel. But was this decision the only reasonable alternative at the time?

French aviation industry experts could have advised the Air Ministry to procure specialized bomber and fighter airplanes. By 1933–1934, it had become clear that fighter engine designs were well on their way to providing better performance than those for bombers. The advent of high-octane fuel, superchargers, and high-performance wing designs increased the single-seat design advantages dramatically. So, why would Cot and Denain choose a design that was obviously on the wane? The answer lies in the pressure brought to bear by the economic and political crisis and by the army. Since Cot and his chief of staff had limited credits to spend on modernization, the army and navy could not allow the aggressive air leaders to procure pursuit or bomber airplanes that were not suited for observation or close air support missions.

Moreover, industry leaders wanted to produce airplanes that appealed to commercial as well as government customers. Investments in fighter designs would lock the industry into a narrow military market, and the 1920s *politique des prototypes* had proved this to be fraught with uncertainty, as the government allowed firms to undertake research and development without ordering enough airframes to permit them to recoup their investments. Thus, leaders in the senior services and in industry held the Air Ministry hostage. Aircraft firms would not build planes unless the government ordered enough to ensure they turned a profit, and the army and navy would support credits only for airplanes that fit their notions of how airpower should perform on the battlefield.

In retrospect, the first generation of BCR aircraft (1934–1935) should have been the last. Instead, French firms continued to make incremental improvements in the designs until 1938. Again, the reasons are obvious. The Armée de l'Air had created an entire organization and training system based on employing formations of BCR-type battle planes. Abandoning the designs and the accompanying service structures probably would have spelled the end of the air force as an institution. The airmen became caught in a trap that forced them to try to perfect a flawed system in the face of increasing evidence that their technological gamble was bankrupt. To recast the air service into one that was better able to meet the Luftwaffe on more or less equal terms would have required great courage and large amounts of political capital. The British managed to pull off such a feat when the minister for the coordination of defense forced RAF leaders to shift procurement emphasis from heavy bombers to fighters and radar. As late as 1938, the French could have attempted a similar technological shift, but the institutional battle lines had ossified to the point that all the Air Ministry could do was attempt an incremental solution to the problem. The result was the D-520—a technological solution that the Armée de l'Air could have used to compete effectively against the Luftwaffe had the war occurred in 1941 or 1942, when sufficient numbers would have appeared in the squadrons.

A second, perhaps more telling turning point occurred in February 1940, when Air Minister La Chambre and Chief of Staff Vuillemin surrendered their service's organizational structure to the army. If there was a single area in which airmen were responsible for the defeat, this was it. French military aviation had operated since 1933 by attempting to follow the principle of concentration of forces. Air leaders had succeeded in gaining the army's acceptance of the necessity of having an air commander to advise army and navy commanders on the proper use of aviation assets. Since 1933, annual exercises and war games had reinforced the concept that only by concentrating scarce airpower resources could commanders expect to achieve maximum effect on the battlefield. Therefore, the decision to change how aviation command and control would function, occurring after the French declaration of war and three months before the Germans attacked, amounted to the air leaders' abandonment of their duty to employ air capabilities correctly.

This decision amplified the shortcomings that the airmen knew existed in their service. The weaknesses of the alerting networks, the poor readiness of the reserves, and the inadequacy of the logistical systems all came under greater stress as the reorganization and later the pressures of combat dismantled the geographic command structure. After seven years of doctrinal development and experimentation that emphasized the operational and strategic utility of airpower, French air leaders allowed the army to force it into a mold that, at best, gave the air service only a tactical role. In their country's hour of greatest need, airmen chose to restrict their vision of the war to the cockpit. This loss of operational vision and the inability to present the unique aviation options to the supreme war council deprived France of one of its most potent weapons.

Finally, the French failed to operate their part of the national defense structure in ways that would lead to effective combat performance. The clear lack of trained, proficient aircrews limited the operational capability of the Armée de l'Air. This stemmed, in part, from a willingness to accept the army's view that the pace and scope of warfare had not changed appreciably since the last war. Airmen chose to ignore lessons they could have learned about airpower from the various small wars of the interwar period. They underestimated the efficiency of fighter airplanes, they overestimated the effectiveness of ground-based air defenses, and, curiously, they overestimated the effectiveness of their own aerial striking capabilities while simultaneously underestimating the same capabilities of their adversaries. These were clear intellectual and operational shortcomings that professional aviators should have avoided.

In the final analysis, the Armée de l'Air deserves a considerable amount of the blame for the German conquest of France in 1940. Was the French air service the primary culprit in the defeat? Was it the subversive element that conspired to open the doors for Germany by purposely failing to do its duty? The answer is no on both counts. French airmen exhibited failings that were similar to those of their surface warfare counterparts in the army and navy. They served honorably in combat, and many of them died attempting to counter forces that were better suited in terms of technology, tactics, organization, and operations.

Analysts seeking to learn lessons from the Armée de l'Air's interwar experience should recognize that political, interservice, and economic pressures; technological constraints; and organizational decisions came together to force hard decisions that military and civilian authorities did not necessarily want. Leaders, however, often have to make suboptimal choices to carry out their duties in the politically charged realm of national defense. Such decisions may reflect the only choices, and they may even be "right" choices. But when a number of suboptimal choices come together in a time of crisis, such as occurred in the 1940 Battle of France, institutions and individuals find themselves at a loss to explain how they could have been so blind.

Suggestions for Further Research

Studies of the interwar French air force in English are rare compared with such studies of the American, British, Japanese, German, and Soviet air arms. Scholars who wish to expand the literature in this field could take several paths, one of which is to concentrate on relations between French aviation industries and Armée de l'Air officials to discern how requirements for new aircraft designs and production evolved. French scholars have examined how labor relations influenced aviation production capabilities within the context of domestic politics, but studies that show how airmen communicated their needs to industry professionals are lacking.

A related area is how the French air arm experimented with new technology and new concepts. My research revealed that significant experiments were conducted to test low-altitude and dive-bombing techniques in the 1930s, but an institutional study of this capability does not exist in French or English. As with many of the issues related to French airpower in the interwar period, the value of such a study may lie in showing how the French failed to apply valuable lessons from their admirable investments in innovative structures rather than how such initiatives led to a more effective fighting force.

Airpower history tends to focus on how significant personalities influenced doctrine, force structure, and operational capabilities. French airpower history is marked by a dearth of such studies. Sabine Jansen's recent biography, *Pierre Cot: Les Pièges de l'antifascisme (1895–1977)* (Paris: Fayard, 2002), is a notable exception. A more complete body of literature would include biographies of Generals Victor Denain, Paul Armengaud, and Joseph Vuillemin, along with Air Minister Guy La Chambre.

On a tactical level, the evolution of interwar paratroop doctrine could reveal how French thinking in the 1930s led to the development of the paratroopers that played such a significant role in conflicts in Indochina and Algeria. Similarly, a blow-by-blow study of the 1940 campaign would contribute significantly to World War II airpower history. Such a study would benefit from an engineering comparison of French, German, and British aircraft, doctrine, tactics, and operational maneuvers.

Finally, historians have analyzed alternative paths for most World War II combatants to assess how the outcome might have changed in light of different decisions. Such a study focusing on a better use of available French air forces would provide a more comprehensive perspective of the opening gambits of the war.

Recommended Reading

In *The Hollow Years: France in the 1930s* (New York: W. W. Norton, 1994), Eugen Weber declares that "French air power influenced policies largely by its absence." See Phillipe Bernard and Henri DuBief, *The Decline of the Third Republic, 1914–1938,* trans. Anthony Forster (Cambridge: Cambridge University Press, 1985). The apocryphal Vuillemin prediction—that the French air force would not last two weeks in a war with Germany—came after he returned from an official visit to that country, where the Luftwaffe showed him a carefully prepared Potemkin village of air fleets and vast modern aircraft manufacturing facilities. The German deception completely demoralized Vuillemin. See Arnaud Teyssier, "Le Général Vuillemin, Chef D'état-Major Général de l'Armée de 'Air, 1938–1939: Un Haut Responsable Militaire Face au Danger Allemand," *Revue historique des armées* 2 (1987), and Patrick Facon, "La Visite du Général Vuillemin en Allemagne (16–21 Août 1938)," *Revue historique des armées* 2 (1982).

L'Aéronautique provided one forum for debating the role of airpower in European security. See Henri Bouché, "Raisons et Moyens d'un Désarmement Aéronautique," *L'Aéronautique* 153 (1932), and "Perspectives de Guerre Aérienne ou de Paix Aéronautique," *L'Aéronautique* 229 (1938).

For one of the best analyses of how German air doctrine proceeded from theory to practice, see Richard Muller, *The German Air War in Russia* (Baltimore: Nautical and Aviation Publishing Company of America, 1992). Also see Richard Muller, "Close Air Support: The German, British, and American Experiences," in *Military Innovation in the Interwar Period,* ed. Williamson Murray and Allan R. Millett (Cambridge: Cambridge University Press, 1996), and Thierry Vivier, "L'Armée de l'Air et la Révolution Technique des Années Trente (1933–1939)," *Revue historique des armées* 1 (1990).

For a provocative view of the French air force and the defeat of 1940, see the following articles by Faris R. Kirkland: "Plans, Pilots, and Politics: French Military Aviation, 1919–1940," in *1998 National Aerospace Conference—The Meaning of Flight in the 20th Century,* ed. J. F. Fleishauer (Dayton, Ohio: Wright State University, 1999), 285–93; "French Air Strength in 1940," *Air Power History* 40, no. 1 (February 1993): 22–31; "The French Air Force in 1940: Was It Defeated by the Luftwaffe, or by Politics?" *Air University Review* 36, no. 6 (September–October 1983): 101–18.

For an excellent analysis of British innovations, see Alan Beyerchen, "From Radio to Radar: Interwar Military Adaptation to Technological Change in Germany, the United Kingdom, and the United States," in *Military Innovation in the Interwar Period,* ed. Williamson Murray and Allan R. Millett (Cambridge: Cambridge University Press, 1996).

For one of the best comparisons of the French methodical battle and German *auftragstaktik* doctrines, see Robert Allan Doughty, *The Breaking Point: Sedan and the Fall of France, 1940* (Hamden, Conn.: Archon Books, 1990), and *The Seeds of Disaster: The Development of French Army Doctrine, 1919–1939* (Hamden, Conn.: Archon Books, 1985).

For an excellent series of essays that capture the similarities and differences among interwar airpower theories in one volume, see Phillip S. Meilinger, ed., *The Paths of Heaven: The Evolution of Airpower Theory* (Maxwell Air Force Base, Ala.: Air University Press, 1997).

For a fascinating study of how the famed military theorist Sir Basil

H. Liddell Hart influenced the British policy of limited liability before the Second World War, see John J. Mearsheimer, *Liddell Hart and the Weight of History,* ed. Robert J. Art and Robert Jervis (Ithaca, N.Y.: Cornell University Press, 1988).

For an excellent revision of the causes of World War I and the relationships among the key participants, see Richard F. Hamilton and Holger H. Herwig, ed., *The Origins of World War I* (Cambridge: Cambridge University Press, 2003), particularly Eugenia C. Kiesling's chapter on France and J. Paul Harris's chapter on Great Britain.

For a brief narrative of the destruction of French medium bombers during the Battle of France, accompanied by excellent photos and aviation art, see Olivier Ledermann and Jean-François Merolle, *Le Sacrifice: Les Breguet 693 de L'Aviation d'Assaut dans la Bataille de France* (Paris: IPMS France, 1994). For an account of how British light and medium bombers fared over the battlefield, see Denis Richards, *The Hardest Victory: RAF Bomber Command in the Second World War* (New York: W. W. Norton, 1994).

Cot identified the failure to solidify the alliance with the Soviets as one of the principal causes of the defeat. See Pierre Cot, *Triumph of Treason: "Contre Nous de la Tyrannie . . ."* (Chicago and New York: Ziff-Davis, 1944). For the most comprehensive biography of Cot, see Sabine Jensen, *Pierre Cot: Un Antifasciste Radical,* ed. Serge Berstein and Pierre Milza, Nouvelles Études Contemporaines (Paris: Fayard, 2002). See also Pierre Cot, *L'Armée de l'Air* (Paris: Grasset, 1939).

A perceptive contemporary view (written in 1940) is that of Marc Bloch, *Strange Defeat: A Statement of Evidence,* trans. Gerard Hopkins (London: Oxford University Press, 1949). Anatole de Monzie, *La Saison des Juges* (Paris: Flammarion, 1943), represents the post-Riom inquest standpoint.

For an account of French foreign policy efforts in the 1930s, see Jean Baptiste Duroselle, *Politique Étranger de la France: La Décadence, 1932–1939* (Paris: Imprimerie nationale, 1979).

For a masterful analysis of the evolution of Soviet policy and its effect on the war's early phase, see Gerhard Weinberg, *A World at Arms:*

A Global History of World War II (Cambridge: Cambridge University Press, 1994).

Vivier has written several comprehensive analyses of the nexus between French airpower and diplomatic efforts with respect to eastern Europe. See Thierry Vivier, "La Coopération Aéronautique Franco-Tchécoslovaque, Janvier 1933–Septembre 1938," *Revue historique des armées* 13 (1993); *La Politique Aéronautique Militaire de la France, Janvier 1933–Septembre 1939* (Paris: L'Harmattan, 1997); and "L'aviation Française en Pologne, Janvier 1936–Septembre 1939," *Revue historique des armées* 193 (1993).

Also see Pascal Vennesson, *Les Chevaliers de L'Air* (Paris: Presses de Sciences Po, 1997); Herrick Chapman, *State Capitalism and Working-Class Radicalism in the French Aircraft Industry* (Berkeley: University of California Press, 1991); and C. Rougeron, "What Chance Has Pursuit? A French View," *Aviation* (April 1936).

World War II started with the use of concrete runways at fixed bases. Later in the war, mobile temporary runways could be created in one to three days. This is Evreaux-Faurville in northern France on 9 July 1944. When an air force has the ability to hit its opponent's bases, the bases become prime targets.

Polish-designed and -built 75 mm antiaircraft guns. In September 1939 there were only forty-four of these guns in service. Courtesy of the Pilsudski Institute of America, New York.

Faculty and staff of the Polish Aviation Service Academy, Deblin. General Ludomil Rayski, commander of the aviation service, is seated fourth from the right; Colonel Stanislaw Ujejski, commandant of the academy, is seated fourth from the left. Michael Peszke's father is standing in the first row, at the extreme right. This photo was taken in 1933, before the introduction of the steel-blue uniform. From the Peszke archives.

Rear view of the PZL 23.

Swiss license-built Morane D-3801 single-seat fighter. The usually astute Swiss bet on France's being the major military airpower in 1939.

A Mig-19 code-named "Farmer" by the Allies. Its internal rocket rack is in the firing position.

To match Western jets, the Soviets supplied their latest equipment, such as this Mig-19, which saw action in the Middle East.

Widely supplied to client air forces, the Mig-21 fighter was a short-range, high-performance machine whose high-speed landing run had to be stopped by a parachute. Dissatisfaction with Soviet arms resulted in the Westernization of Arab air forces.

The Fokker D7 of 1917 was probably the best German fighter of the war. Given restricted engine production, the emphasis was on aerodynamics and a defensive strategy.

The other great German fighter, also of Dutch design, was the tiny Fokker triplane.

Opponents included American Captain Eddie Rickenbacker, shown with his French Spad S-13.

General Giulio Douhet, the theorist who tried to make Italy a major airpower by advocating the bombing of others beyond the Alps. Italy could not afford his ideas.

The Savola-Marchetta SM-79 multipurpose bomber (here carrying a torpedo) scoured the Mediterranean.

The Fiat CR.32 of 1931 was already outdated and outclassed at the outbreak of war on 10 June 1940.

The Macchi 2002, Macchi 2003, Fiat G.55, and Reggiani 2003 were linked by use of the German DB-605 engine. These aircraft first appeared in 1942, with the 2003 going into service in April 1943. It was judged one of the best single-seaters of the war.

CHAPTER THREE

The Arab Air Forces

Robin Higham

The modern Arab Middle East is difficult to analyze because one umbrella term covers a variety of nations, states, religions, economies, developments, and political goals. Trying to do so is no more productive than trying to analyze the West, but at least the West is better understood in Europe and America.

The differences among the Arab states' inherited legacies and their cultures, and the question of how "backward" they are, may be puzzling to a nonspecialist. Nevertheless, it is necessary from a diplomatic and a military viewpoint to assess the Arab capability for modern war. One of the tools for understanding the situation is C. P. Snow's *The Two Cultures,* referring to the intellectual literati on the one hand and engineers on the other. There is no denying that in the past the Muslim world contributed to art, literature, and mathematics and was fundamental to the West's progress from the Renaissance to the present. But in practice, the Arab world (as elsewhere) is highly political, to the extent that the armed forces, notably the armies, have been instruments of domestic, not foreign, policy. Moreover, until the advent of Israel (officially in 1948), the sparseness of the population meant that there were few borders in dispute—in spite of the arbitrariness of the Ottoman Boundary Commission's actions after the peace of 1920—until the Arab states gained independence from colonial powers and could start building armies.

The politicized armies would not have faced manpower problems if their only functions were to build nations and to keep the domestic peace; what changed the picture is the multitude of international wars they have been forced to fight against modern opponents. These wars have compelled the Arab armies to modernize, to become highly technological, and to scrape the proverbial barrel for suitable manpower to run the new systems and address the problem of training and motivation. This, in turn, has required them to undertake educational programs to obtain the manpower (and, where permitted, woman power) needed. It has also forced them, at times, to embrace Western values or structures in order to match their Western opponents' capabilities. This has caused a certain degree of what could be called culture shock, or cultural strain. Although they have often been the leaders of modernization in their part of the world, the imperfect mirror-imaging of their Western opponents has led outside observers and others to characterize the Arab armed forces as backward (a condition recognized in the United Nations' 2003 *Arab Human Development Report*).

Another way of looking at the situation is by taking a historical overview of the development of civilizations. Although the Industrial Revolution started in Britain in about 1760, France was still a peasant economy in 1939; the difference between the two lay in the inheritance of land and the growth of industry and railways. That process had not yet come to the Middle East in Ottoman days (1453–1923). So comparatively speaking, major parts of the Arab world were still backward. But as Snow noted in 1961, those countries could catch up in a generation.

In Egypt's case, it was galvanized by the 1956 and 1967 wars to modernize its armed forces, if not its society. And given the demographic, political, and pecuniary issues that constrained the process, Anwar Sadat and then Hosni Mubarak elected to concentrate on air and armor.

Whether modern states in the Middle East have been hampered by military dictatorships depends on one's definition of the term and on the nations themselves and their political, economic, and national security situations. Education, democracy, and technical and technological progress must be related to internal stability and external threats. Both Jordan and Pakistan have had to handle volatile influxes of refugees—Palestinians from Israel and Afghans, respectively. Both Israel and Afghanistan are engaged in guerrilla warfare and illicit transborder traffic, and in Pakistan's case, air incursions put a strain on both border and internal security.

As described later, the mismatch between the Israeli and Arab air forces from 1948 to 1973 was critical to events, both military and political, in the region. In the thirty years since the Ramadan–Yom Kippur War, the Egyptian and Israeli air forces have reached parity, if not in numbers then at least in will and skill. Neither group desires a renewal of war, and neither has anything to gain from one. In the north, there has also been relative calm between Israel and Syria, although this is primarily because the Syrians—who had to defend the Bekaa Valley, where the Palestinians maintained bases—were badly beaten in the air. Still, at the state-to-state level, war is not wanted there either, but a continuing source of friction is the Palestine Liberation Organization's militant wings and Hamas and its allies. In contrast, the Iraq–Iran stalemate of 1979–1989 (the first Gulf War) was between two of the more advanced Middle Eastern states with better-educated populations.

The young pilots of 1973 have now passed through their professional careers and retired or long ago left the service. The matriarchal matrix (grandmother power) has died, faded, or been deflected by the benefits of peace and wealth. The patterns I laid out in "Ending Enmities"—that conflicts fade as states become joined in unions—are being followed. For one reason, social and economic demands have taken precedence as populations seek a higher comfort zone, and no one wants a conquest that will only create another huge minority problem, such as Israel. In the areas where air forces existed and were eliminated—Iraq and Afghanistan—tribalism, religion, and nationalism will have to be melded into viable states before peace can be achieved.

The defeat of air forces in the Middle East and Southwest Asia as a whole has led to either a parity stalemate or to the elimination of the air arm as an offensive weapon, at least by the "loser states."

The Theater

The physical terrain is typified by aridity and a lack of cover. In terms of climate, there is great heat in summer and intense cold in the mountains in winter. Sandstorms can also affect operations. There is little cloud cover, and the weather and visibility favor flying. In 2003, 70 percent of Iraq was cloud-free 30 percent of the time; 65 percent of all sorties canceled because of weather occurred on three sandstorm days. The theater has been characterized by high wastage and consumption of material for several reasons. Pressures of operations forced Muslim pilots through the training process too quickly, placing them at a disadvantage in action. Although accident rates and causes are unknown, we can interpolate from other forces that losses from pilot error and mechanical failures have been substantial. Sand and dust are both an operational and a maintenance headache.

Central to an understanding of airpower in the Middle East is the rise of the Israeli air force (IAF) as the most successful air force in the region. It is impelled by the necessities of national defense. Although Israel has an educated (92 percent literate), industrious population, an indigenous aircraft industry, and strong financial and other ties with the United States, it cannot afford long wars and so has to keep its force operational, flexible, and, as much as possible, compatible. Given the absolute necessity of maintaining air superiority, the IAF has sought the best, simplest, and most reliable equipment manned by superbly trained personnel. The Egyptian, Syrian, and Jordanian air arms are affected most directly by the actions of the IAF, but it has also impinged on the Saudi Arabian, Iraqi, Libyan, Yemeni, and Algerian forces, not to mention at times the Soviet, British, French, and, indirectly, even the Iranian and Pakistani forces.

Throughout the Middle East there is a strong legacy of British establishments, organizations, operational procedures, and training. Ranks and insignia, even brevets, demonstrate this influence from Egypt to Pakistan. Notable in the Arab world has been the shift from British aircraft. Some countries have received U.S. aid and thus American aircraft, but Soviet machines are more common, especially because the authoritarian nature of the USSR made policy decisions easier to make than in the United States or Britain, where democratic forces had to be considered, or in France.

Complicating the Arab thought process are the shifting relationships among the Arab states themselves—sometimes allies or even united, and sometimes enemies—and their determination that Israel is the devil incarnate and that, as its sponsor, the United States is tarred with the same brush. This latter was a convenient straw man for the unwillingness of some Muslim states to buy and learn to use Western aircraft and weapons systems, in some cases because they could not afford them, and in others because anti-Semitism and anti-Americanism were fuses for noisy public demonstrations by people who lacked political power and were not admitted to policy councils.

Interestingly, there are parallels between the loser Arab air forces and the South in the American Civil War a century earlier. The Middle East has much space but little infrastructure; an ill-developed transportation system, except for oil pipelines; and societies divided between the fellahin and the aristocratic, dictatorial upper class, with a small, powerless, and hence ineffective middle class. On the whole, there has been a lack of understanding of the nature of modern war and a conflating of hardware with viable power. As a result, in wars of any duration, the Muslim states have been forced to fall back on guerrilla operations because they have been unable to grasp victory in short, hard, highly consumptive wars.

Another factor affecting the Arab air forces' history of defeat is the predominance of the Israeli side of the story (aided by Arab secretiveness) and world admiration for the IAF's flexible use of airpower from a central base. Seen as the world's most continuously experienced air arm, the IAF enjoys enormous prestige.

But in the thirty years since the Ramadan–Yom Kippur War of 1973, the Egyptian air force (EAF) has gained parity with the IAF in terms of skill, morale, and weapons, and its close association with the West has enabled a more balanced assessment of its strengths and weaknesses. Also crucial to a reassessment of the Arab–Israeli wars are the doubts cast on the hijacking of history. IAF claims and the now guardedly admitted Egyptian losses simply do not add up, even when dividing claims by the standard three to get closer to reality.

Whereas the IAF has been able to exploit the flexibility of airpower and of fighter-bombers from a central position on interior lines, combined with the unity of command of the whole Israeli Defense Forces (IDF), the Arabs have suffered from exterior lines, divided commands, and different, even conflicting objectives, although there has been much commonality of equipment.

In the latter twentieth century, air defenses came to be an integral part of air force strategies, and to counter their effectiveness, all sides have employed both suppression of enemy air defenses (SEAD), including ground operations, and drones or unmanned air vehicles (UAVs). The importance of airfields, airfield defenses (especially radar), and communications increased enormously from 1945 to the present, including airborne warning and control systems (AWACs). Unfortunately, in most of these areas, the Arab air forces are behind, in spite of oil revenues. Even in the case of Iraq in 1990–1991, the massive Soviet-supplied air defense system was quickly dismantled by Coalition SEAD techniques and tactics.

The clearest example of the ramifications of arms supply came in the 1973 Yom Kippur War, when both the Soviet Union and the United States organized combined resupply and diplomatic efforts to safeguard their clients. Complicating such moves has been the democratic process in the United States, lobbying in France against sales to Israel, and surplus production in the Soviet Union combined with fewer inhibitions on the export of defense secrets. U.S. concerns and congressional sensitivities forced Israel to abandon the U.S.-engined Kifir, which was then sold to South Africa. In this vast cockpit of dilemmas, a moralistic foreign policy can easily backfire in the search for others in the souk.

At the same time, the air war has pitted generations of the latest fighters against one another. The IAF started with British and French jets in 1956, followed by American planes when the French banned further sales. The opponents have fielded Mig and Sukhoi generations all along to face the A-4, F-4, F-15, and F-16 planes in IAF service. Necessary enhancements have been radar, avionics, and missiles. The Iraqi air force acquired Dassault Mirages, so in 1991, the French AF Mirages were sent home before the Gulf War to avoid "amicide."

History

After 1948, both Israel and Egypt spent the next seven years buying new aircraft, training, and setting up. The EAF became attached to the Soviet bloc in the Cold War and received quantities of IL-28 bombers and Mig-15 and Mig-17 fighters starting in 1955 to replace its aging first-generation British Meteors and Vampires. The IAF began to receive Anglo-French support. In the meantime, the British had withdrawn from the Suez Canal Zone, with an agreement to keep war stocks and defense of the canal in their hands. But when the West refused to supply Egypt with arms and funds for the Aswan High Dam, Nasser appealed to the Soviet Union, which provided arms via Czechoslovakia. He then nationalized the canal on 28 July 1956.

On 31 October the British Royal Air Force (RAF) attacked Egyptian airfields but did little damage, while the Armeé de l'Air (French air force) supported the IDF. After that, there was no more air combat; the EAF had been neutralized. The Allies lost 10 aircraft, and the EAF lost perhaps 250. As in 1949, the EAF should have learned that it lacked air defenses, modern aircraft, and training. On 5 November the Cold War intervened with a Russian threat to "nuke" London and Paris. The United States, through the International Monetary Fund, agreed to guarantee $500 million of support to Egypt if there was a cease-fire. All troops left Egypt by 22 December 1956. The war increased Anglo-French distrust and weakened the Anglo-American relationship. Oil to the West now had to go around Africa because the Suez Canal was blocked and the Syrian pipeline broken.

The war showed both sides the need for full modernization, and it convinced the EAF of the need to streamline its command system. In

1956 the IAF had a sortie rate of about three to four a day, versus the EAF's one.

Egypt and Syria, the principal Israeli opponents, had a combined population in 1990 of 68 million, roughly 10 percent of which lived in either Cairo or Damascus. However, both countries' populations were evenly divided between urban and rural. Egypt's literacy rate was only 48 percent, compared with Syria's 64 percent. While 99 percent of Egyptians lived on 4 percent of the land, Syrians were much more evenly distributed.

Egypt was an agrarian society spread along the Nile Valley and in the Nile Delta. It was, in the 1960s, still heavily stratified with rich families, some of whom were military and some of whom, especially in Alexandria, were Greek. There was an urban middle class and rural landlords. In addition to the urban poor, there were the landless villagers, who made up at least 50 percent of the population. Intellectually, Egypt was way behind in terms of the technocrats and professionals and the education and training needed for war against the advanced, modern IAF, as indicated by the 1966 Egyptian Higher Commission for the Liquidation of Feudalism.

In 1957, upon reviewing the 1956 Sinai War, President Nasser designed a grand strategy to confront Israel, deter further Western aggression, and enable Egypt to lead the Pan-Arab movement. So the EAF was totally reequipped and heavily expanded, along with mobile and armored units. This was Egypt's first carefully considered grand military strategy.

A mass of new Soviet equipment and some 650 Soviet and Czech pilots and technicians arrived, translated manuals into Arabic, and pushed for the EAF to be modeled along Soviet lines. The few new types of aircraft simplified training and maintenance. New Egyptian personnel were recruited, and large numbers were sent to the Eastern bloc to be trained, including selected staff officers who were dispatched to Moscow. EAF morale soared.

Joined with Syria and Yemen, the United Arab Republic Air Force (UARAF) had a paper strength of thirty frontline squadrons, but these had limited experience, little technical expertise, and shoddy training. British-trained personnel were rifted; the new concentration was on flying safely (to cut the high accident rate), maintaining aircraft, and obeying commands. These restrictions would be costly in the future. British influence remained in the counterstrike philosophy for use of the twin-engine Soviet-built IL-28s.

From the late 1950s, the Egyptian government aggressively pursued an expansion of the indigenous defense industry. By 1961, the EAF had eighty Mig-19 Farmer supersonic jet fighters, followed a few months later by the Mach 2 Mig-21 Fishbed interceptor. The latter carried only limited fuel.

Once the Soviets began supplying the Arab air forces, a fundamental problem arose: Russian air defenses had been developed to deal with high-flying Allied heavy bombers, such as the B-47, B-58, B-52, Handley-Page Victor, and Avro Vulcan—targets for interceptors—while an array of surface-to-air missiles (SAMs) and light antiaircraft weapons were aimed at ground-support aircraft at low levels. Opposed to the Soviet largesse were Israeli and eventually U.S. Air Force and RAF fighter forces intent on winning air superiority by dogfighting at various altitudes and by neutralizing air bases.

The Middle East, especially in the Arab–Israeli theater, was the prime source of ongoing intelligence on both sides, including the lessons from Vietnam.

Tensions in the Middle East resulted in constant pressure to update equipment, especially electronics and missiles; this pressure was countered by the reluctance of producing countries' politicians to risk the unauthorized transfer of advanced technology to Cold War opponents. The politics of nationalism, religion, and authoritarianism were also factors in states whose relationships resembled a bunch of magnets hanging from a common point with north poles uppermost.

By the time of the 1967 Six-Day War, the EAF–UARAF was confident but untested. Experience had been diluted by expansion. Serviceability rates were at the 60 to 80 percent level, about the same as for the RAF in 1943–1945. Command and control systems hampered effectiveness. Moreover, the Soviet welded-wing formations, when coupled with limited training, made the Egyptians, Syrians, and Iraqis no match for the fluid pair tactics of Western pilots.

In April and May 1967 the United Arab Republic (UAR) forces mobilized in response to the IAF victory over the Golan. Nasser made aggressive statements, and the Arabs blundered into war. The EAF lacked a theory of air combat and had been trained for high-altitude work, not to meet the IAF, which had trained against captured Migs, at low levels.

On 6 June the overconfident UAR got its comeuppance. The UARAF was outnumbered and destroyed, but not without a valiant fight. The EAF admitted the loss of 43 aircraft, versus the IAF claim of 336 shot down or destroyed. The 1967 war was an old-style 1939–1945 battle. Soon thereafter modern war came to the Middle East.

Nasser responded to the 1967 defeat by firing all the high commanders (failed Egyptian officers were tried and jailed), appointing a new chief of the air staff, rehabilitating personnel, assessing losses, reorganizing combat units, and developing a rebuilding plan. The Russian chief of staff visited Cairo in late June and refused to provide new arms until the EAF was thoroughly trained. The Soviets sent massive antiaircraft support to defend EAF bases. Public confidence in the military was not restored until after the 1973 Ramadan–Yom Kippur War.

By autumn, the aircraftless pilots were remounted in fresh Migs supplied by the USSR, were learning to fly at low levels, and were training intensively. They also struck Israeli positions. Before the end of 1967 the EAF was more than 80 percent refurbished and its air bases hardened.

By 1968, Egyptian resolve led to the War of Attrition, when the Egyptian Air Defence Force (EADF) was created on the Soviet model. The lessons of 1967, as in Britain in 1940, saw the coordination of antiaircraft defenses, fighters, and observation posts to detect low-level penetrations. By 1969, the EAF was crossing the Sinai with escorted strike forces. But the new ex-U.S. IAF A-4s and F-4s had both higher speeds and better electronics, as did the older Mirage IIIs. In late 1969 the IAF launched an offensive against the air defenses that inflicted casualties and lowered morale. In early 1970 F-4s and A-4s, evading the defending guns, missiles, and Migs, struck deep around Cairo, forcing a shift in the defenses.

Soon thereafter, Soviet combat forces deployed to Egypt and went on the offensive; engagements were intense. When Soviet Mig flights were ambushed, the EAF was delighted—Soviet pilots were not so superior after all. This phase ended with a cease-fire brokered by the United States and the USSR. The EAF had proved itself in action in spite of the IAF, with the loss of 109 aircraft and 90 aircrew in combat and in operational accidents and another 50 machines in training accidents. Seventeen IAF planes had been shot down.

The 1970s began with both sides changing tactics. Egypt revised its air defenses, and the IAF used its air superiority to attack military targets around Cairo to bring the conflict to the attention of the Egyptian public. What it did was to alert the Soviets to the danger to their client. Russia at once sent 150 volunteer pilots and more than five squadrons of Mig-21s and batteries of the low-level SA-3 SAMs and ZSU-23 radar-directed quad 30 mm guns to provide defensive belts along the Suez Canal and around Egyptian cities. By April, the IAF had stopped penetrations (averaging 800 a month) in order to avoid a confrontation. Israel itself was covered by high-altitude Soviet PRU sorties. Then the EAF began a series of strikes across the canal with SU-7s. To stop these, the IAF went on the offensive again but met increasing Russian resistance in the air, resulting in mounting losses in dogfights as well. By May, some 15,000 Soviet advisers controlled six airfields and thirty-seven SA-3 sites in Egypt. Casualties rose as the IAF laid traps to bounce defending Migs. The more disciplined Egyptians and the more reckless Syrians had lost 113 aircraft to the IAF's 35. Diplomacy called a halt to the War of Attrition on 8 August 1970.

Nasser died in September 1970 and was succeeded by Vice President Anwar Sadat, who gave the EAF an important role. An immediate study showed that the EAF and EADF working together could be successful. At the same time, Hebrew-speaking Egyptians were recruited for intelligence work, and secure communications were installed to keep pilots informed. A new program of air base protection put the EAF into the world's lead.

Soviet refusal to share Mig-25 Foxbat reconnaissance photos of Israel worsened relations with Egypt in 1971 and led to the Soviets' expulsion in 1972 over their refusal to supply offensive weapons. Sadat made a foreign-policy decision and refused to keep Russian equipment under Soviet control. The USSR left behind the aircraft and weapons but took the electronics. Egypt retained the new SA-6 SAMs and ZSU-23 guns and the new Su-20 tactical aircraft, to which EAF Su-7 pilots easily converted.

In 1972 Air Vice-Marshal Hosni Mubarak became air force commander in chief and principal planner for an attack on Israel. Quiet discussions for the purchase of European weapons were initiated, as well as talks with Syria on joint operations with Libya. Training on new Mirages was led by Egyptian, Libyan, and Pakistani air force instructors. Sadat persuaded a reluctant military to go to war for limited objectives to achieve peace with Israel. Surprise was vital to seize the Bar-Lev Line and the Jiddi Pass and then dig in to defeat IDF counterattacks.

On the eve of the Ramadan War, the EAF had 23,000 personnel, 730 pilots, and 800 aircraft, as well as Libyan, Iraqi, and Algerian squadrons to supplement the Syrian and Egyptian forces. The EADF had 100,000 additional men.

The 1956 and 1967 wars had cost the EAF a few pilots and had sunk morale. So in 1973 the EAF plan was to avoid close encounters and the battle for air superiority and to concentrate instead on sudden ground attacks.

Surprise was achieved by a disinformation campaign aided by Israeli intelligence's misinterpretations. Alerted on 6 October, Prime Minister Golda Meir refused to allow preemptive strikes because of U.S. pressure. The IDF faced air forces in hardened shelters and Arab ground forces that were defended by a thick air-defensive belt. All Arab aircraft carried identification, friend or foe (IFF) systems, and control centers were joint EAF–EADF operations. The 200-machine Arab first strike seemed like a maneuver. IAF airfields were cratered by Egyptian-made dibbler bombs and strafed by well-practiced professionals.

IAF attacks on Egypt were beaten off by the combined EAF–EADF defenses, using finger-four tactics in the air. They were happy to avenge the 1967 defeat.

On the Day of Atonement the Egyptians created eleven bridgeheads across the Suez Canal, each four miles wide. The surprised IAF scrambled fighters, which suffered heavy losses from SAMs. The EAF struck three airfields in the Sinai, as well as radar sites and Hawk missile launchers. By 1600 hours the IAF had lost thirty machines, affecting pilot morale, but the IAF was still engaging in an organized response, mainly against Egyptian aircraft operating beyond air defense cover.

In the north the Syrians seized the Golan. Missiles or their threat made IAF attacks imprecise when done from altitude, and the IAF A-4s faced masses of missiles. So by feints, the IAF "drained" the Syrian missile batteries dry by noon on 8 October. By 16 October, Israel had seized the initiative on both the Golan and Suez–Sinai fronts.

As the battles raged, resupply became critical. Both the Soviets and the Americans aided their clients. By the fourth day, the Israelis had pushed the Syrians back beyond their start line and were attacking their airfields. IAF losses fell due to new tactics and U.S.-supplied countermeasure equipment, including chaff. Helicopters spotted SAM launch sites, which were then targeted by fighters. On the fifth day, Syria was out of the war, and Arab losses mounted. Jordan entered the war, but it was too late. The IAF attacked EAF bases to deter close support. The Israelis then followed up by crossing the canal between the 2nd and 3rd Egyptian Armies and aiming at the SAM batteries. The EAF caused high casualties among the attackers. In this last phase the IDF captured Egyptian airfields and used them as resupply bases. Still, the EAF fought valiantly both in ground attacks on the pontoon bridges and with Mig-21s in air defense, some perhaps flown by North Koreans. On 24 October a second cease-fire held, and fighting ended after nineteen days.

This war reinforced the lessons of the War of Attrition. Coordination, training, numbers, and aggressiveness blunted the IAF's efforts and cost it some 110 aircraft, compared with the EAF's loss of 50 to 60 to friendly fire, in part due to incompatible IFF on the Syrian front, and perhaps another 120 (the figures were still secret in 2003) in combat. In addition, about 100 pilots, most of them senior airmen, were lost.

During the war, the Soviets flew 6,000 metric tons of supplies to Egypt and an equal amount to Syria and Iraq, while the United States flew 23,000 metric tons into Israel along with thirty-five new F-4s and fifty A-4s. By 1974, the IAF was at 125 percent of its early 1973 strength, but it had been chastened.

The IAF loss was 1.27 percent of sorties flown; it took roughly fifty-five SA-6s to get a kill. The IAF had retained air superiority, and its ultimate losses were only 115 planes in 11,233 sorties, versus the Egyptian claim of 328 kills, which, when divided by the usual three, is about right. The EAF is estimated to have lost 247 planes, the Syrians 179, and the Iraqis 21, for an Arab total of 447.

After the war ended on 24 October, the immediate impact was an Arab political and economic mobilization against the West, including an oil embargo. Egypt and Syria could negotiate from strength. The EAF had suffered heavy losses but emerged self-confident and with higher morale. It renewed relations with the United States, the United Kingdom, and France.

Both sides drew lessons from this contest, which was contemporary with the Vietnam War. And both sides supplied their sponsors with valuable experience and intelligence. It was obvious that winning had to be done beyond the defensive belts, which meant offensive actions. Moreover, consumption and attrition were such that a strategic resupply airlift was essential if a war went beyond seven days' duration. For the first time, the Arab forces had performed well from planning through execution, even if they had not yet learned the necessary wartime flexibility.

The Arab embargo immediately raised the price of oil internationally. This, however, had a beneficial if unintended impact in the West, because it made the population fuel conscious and set the pace for far more efficient aircraft and auto engines. So the long-range effect was both economic and political, in that peace came to the Egyptian–Israeli border and led to negotiated IDF withdrawals.

The 1976 Syrian intervention in Lebanon, with an armored division supported by Su-7s and Mig-21s in ground attack roles, enabled the Bekaa Valley to become a hotbed of anti-Israeli activity, watched by IAF RF-4E overflights. Palestine Liberation Organization (PLO) action led to Israeli intervention in 1978, which included cluster-bombing of guerrilla targets until the thrust was stopped by the creation of the United Nations Interim Force in Lebanon (UNIFIL).

The historic September 1978 Israeli–Egyptian Camp David accords led to the Egyptian–Israeli Peace Treaty of April 1979. However, the PLO set out to disrupt it by attacking from south Lebanon, and the IDF–IAF retaliated, supporting the Lebanese Christian Militia.

Overflights continued, and the Syrian Arab Air Force lost a Mig-25 while attacking the IAF. Tensions escalated, and a second Israeli invasion occurred after an attempt on the life of Israel's ambassador in London. On 4 June 1982 the IAF made multiple raids into Lebanon, followed by ground forces the next day. In all, after a week's action, the Syrians had lost about twenty-six Migs and the IAF had lost thirteen aircraft. The Israelis then laid siege to Beirut. About 15,000 Lebanese died before it was lifted on 12 July. The IAF continued to monitor the Bekaa Valley, losing a special high-altitude camera RF-4E on 24 July. In the aftermath, eleven Soviet engineers who were trying to dismantle the aircraft were killed when an IAF strike destroyed it.

After the 1973 war, Sadat faced economic challenges that forced a reduction in military spending while encouraging foreign trade and investment. After the peace treaty was signed in 1979, U.S. aid poured in. Meanwhile, in 1973 Soviet aid dried up.

Training was intensified, incorporating the lessons of the 1973 war. By the 1980s, EAF officers were participating in exchange programs in the United States, France, and India. A trickle of new Soviet aircraft arrived, but their spares dried up by the 1980s; this reduced fighter pilots to ten to fifteen hours of flying a month. In 1974 Mirages began to join the EAF as the policy changed to permit purchases from multiple suppliers.

In 1979 the EAF received thirty-five Phantom F-4Es, placing the EAF on an equal footing with the IAF. But the new Western equipment placed a strain on Egyptian technical manpower, and new schools had to be opened. The shortage of spares was remedied by Egyptian-made parts and by those imported from China for the Migs, followed by the import of Chinese F-6s (a Mig-19 derivative). Then in 1980 the United States permitted the purchase of F-15 and F-16 fighters.

On 6 October 1981 Sadat was assassinated. Vice President Hosni Mubarak, an airman, succeeded him. Mubarak was committed to modernizing the armed forces and to developing local high-technology and aerospace industries.

The Libyan threat (1977), unrest in the Sudan, the 1982 defeat of Syria, low oil prices, and terrorism after 1985 did not make Egypt secure. Although Egypt had supported Iraq in the war against Iran, it joined the coalition to recover Kuwait in 1990.

From 1980 to at least 1994, air force modernization proceeded as fast as political and economic factors would allow. Five-year defense plans started in 1983 when joint maneuvers with the United States took place. From these, the EAF learned much about U.S. technology, which it incorporated into its F-16s, French Mirages, and missiles.

By the 1990s, the EAF was a modern, efficient, self-respecting air force facing the problems of a long peace.

Second Gulf War, 1991

Saddam Hussein invaded Kuwait, which he claimed as part of Iraq, on 2 August 1990. The postinvasion buildup created the fear that Iraq would proceed into the Arabian peninsula. Five days later, President George H. W. Bush began to deploy Operation Desert Shield to defend Saudi Arabia. As had become the custom since 1945, two carriers were the first to arrive on the scene with their well-equipped supporting battle groups. They quickly joined the Saudi air force and the remnants of the Kuwaiti aid force. Concurrently, U.S. Air Force squadrons in the United States began to deploy within twenty-four hours, and by 12 August, five fighter squadrons were ready to scramble from bases in Saudi Arabia.

Thirty-five days after the invasion of Kuwait, the Coalition (United States, Great Britain, France, Kuwait, Saudi Arabia) had more aircraft in the theater than existed in the Iraqi air force itself. As Saddam appeared unlikely to withdraw, the United Nations approved the use of force if necessary to accomplish this. The Coalition command began to plan the opening air campaign while the rest of the forces assembled in eighteenth-century style to await the battle. There were to be three phases ahead of the ground offensive. By D-day on 17 January 1991, there were 2,430 Coalition aircraft available.

Late on 16 January 1991 the strike packages were on their way to deliver a bomb a minute in daylight all over Iraq. The Iraqi air force rose to the defense but succeeded only in shooting down one of its own Mig-29s. This was the first of thirty-five kills during the war from the sixth largest air force in the world. Even though the Iraqi air force barely got off the ground, the skies over Iraq were hostile with the world's densest air defenses, and this Soviet system was attacked by SEAD tactics to ensure Coalition air superiority. This meant eliminating some 9,500 antiaircraft guns, 17,000 SAMs, and high-tech early-warning systems linked to control radars. The Iraqi air force had 750 combat aircraft on twenty-four main bases and thirty satellites, so one of the major purposes of the Coalition's phase 1 was to destroy this counterthreat. Centrally controlled, the Iraqi air force was blind when its air control system was destroyed by sixty laser-guided bombs and its higher direction was forced underground. By day two, most SAMs were unguided, and antiaircraft defenses could not reach above 10,000 feet, so strike packages flew high.

The destruction of the Iraqi air force was rapidly accomplished, and by the ninth day, it was no longer a factor. Its runways were cratered, and its 600 hardened shelters had been penetrated. The Iraqi solution was to fly 122 of its remaining aircraft to Iran (Iraq's recent enemy, but neutral in this conflict), where they were interned and kept after the war. The result was that by the 2003 war the Iraqi air force had only sixty-nine combat aircraft, as no one would resupply them. An attempt to use Tu-16s to attack Saudi Arabia biologically was destroyed on the ground by intelligence coupled with precise strikes. Whereas in World

War II it took 500 sorties and many enemy civilian casualties to knock out a target, half a century later, it required only one.

By the time phase 3 got under way, the Iraqi army was on its own. The Iraqi air force was no longer there to defend it.

Third Gulf War, 2003

The most recent of the Middle Eastern wars, begun in 2003, is the Third Gulf War, or the War for Iraqi Freedom. This was brought on by the 9/11 terrorist attacks in the United States and the subsequent war on terrorism and al Qaeda, to which, rightly or wrongly, Saddam was linked. Besides being brutal, cynical, and inhumane, Saddam was thought to have both chemical and biological weapons and nuclear potential. He had long proved himself a manipulative and untrustworthy dictator. As it turned out, in 2003 he was also full of bluster.

In March 2003 the United States and Britain invaded Iraq with simultaneous air and ground attacks. The whole campaign was based on wide-ranging, meticulously gathered intelligence, which nevertheless suffered from the inherent danger of disinformation from enemy sources and overzealous belief among Coalition officers.

In the first ten days, the Iraqi air force flew fewer than 120 sorties a day, compared with the Coalition's total of 35,000 plus. The war emphasized the lessons learned in 1991, especially the need to cut down the time between the receipt of intelligence and the launch of a weapon. The complete multisensed command and control could download from Joint Stars, AWACs, and E-2s to the combined Air Operations Center to analyze and decide what to hit. The Iraqi air force lacked such sophistication.

With 70 percent veteran strike crews, the Coalition quickly claimed air superiority, and Baghdad fell in twenty-two days. In Operation Iraqi Freedom (Gulf War III) the Coalition deployed roughly 100,000 airmen (American, British, Canadian, and Australian) against 20,000 Iraqi personnel.

But these incidents do not help explain the collapse of the Iraqi air force and its less than 325 aircraft. Even the sandstorms were not their friend. It seems certain, however, that the no-fly zones prevented extensive training; in addition, many crews were either in Iran or "retrained" upon their return, and morale was low. Further, the Coalition's totally changed tactics to eliminate softening up achieved surprise.

Why was the Iraqi air force so thoroughly defeated in spite of its immense and powerful Soviet-supplied air defenses? The answers are not only the obvious ones—that the Coalition had better planning and execution, equipment, and munitions—but also the Coalition's edge in mental attitude and flexibility, plus an immune command system impregnated with trust. Saddam's airmen did not fare well in a medium in which trust in the system and its senior officers had to be combined with the ability to use initiative. And in part, that was inherited from the Soviet ethos.

Conclusion

The defeat of air forces in the Middle East has often pitted Western technology and training (including, since 1967, sophisticated electronics) against both British legacies and Soviet arms and advisers.

It is notable that due to cost and effectiveness, air action in the area has increasingly been dominated by multimission fighter aircraft, even for ground cooperation. Also notable is the almost total absence of grand strategic air activity and the dominance of tactical airpower, with the seizure of air superiority usually being achieved in a matter of days. Finally, the combined economic and logistical constraints on both sides, except for the Soviets in Afghanistan, have dictated truly limited wars.

The defeat of Arab air forces has often been self-inflicted. In the Arab–Israeli wars, the Arabs' objective has been the destruction of the Jewish state, but they have ended up the losers, unable to accomplish their goal. Arab air forces have consistently been defeated by better trained, more flexible, and centrally controlled opponents.

Because of the Cold War (1947–1991), both sides were supplied with comparable equipment, so it was the human element that counted.

Though their oil revenues could have allowed the Muslims to keep up technologically, especially in electronics, the leadership preferred to spend in other areas. Apart from any religious biases and revolutionary movements, the influence of the clearly Soviet system affected not so much the supply of weapons as the mentality to use them effectively and the willingness to understand the importance of air superiority.

The defeats of the Arab air forces—Egyptian, Syrian, Jordanian, and Iraqi—have been due to a multiplicity of internal and external causes. Although the lack of indigenous aircraft industries, in contrast to Israel, is a basic shortcoming, the paucity of investment in the air forces themselves, especially in training, has also been fatal. Moreover, the Arab air forces have failed to focus on the essential of air superiority—enemy air bases. As a result, they have wasted and consumed a plentiful supply of material, and maintenance is an open question. The latter has also stemmed from an unwillingness or inability to create clear doctrines related to the defense of the nation, a policy that has to be set first in grand strategy.

The only study of a particular Arab air force is Nordeen and Nicolle's of the EAF. It enables a comparison with the multivolume official Pakistani work. What emerges is a picture of the maturity of a civilization influencing the effectiveness of the organization. Even Egypt, the largest Arab state, has taken nearly eighty years and many losses to develop a modern air arm capable by 2003 of standing up to its Israeli counterpart. Why?

The answer lies in the social, religious, economic, and class nature of the societies. Egyptian society has been a placid one consisting of fellahin and a small ruling elite of Turkish, Circassian, and Albanian background since the Ottoman days, which ended only in 1880 with the British occupation. As a result, when the Royal Egyptian Air Force was founded in 1933, it emulated the RAF. But with only a small middle class that was opposed to the idea of a military career, there was no large pool of candidates to become officers or noncommissioned officers (NCOs). The average Egyptian in the 1950s could carry out a detailed plan sent down from the top, could fight stubbornly on defense, but was lost if there was no plan or the unexpected happened. This was true and remains so for other Arab air forces, a number of which employ foreign professionals as pilots and mechanics. But this has been changing.

What the past half century has shown is that the EAF's pilot elite has learned from its losses, not all of whom were killed or captured. In the air they have been forced to learn to dogfight and to abandon the Soviet-style welded-wing formations in favor of more flexible tactics; they have encouraged initiative and flexibility by raising training standards on top of operational experience. The EAF pilot elite has become a professional force, as has that of the Royal Saudi Air Force.

But problems remain. At the top, cronyism and patronage long blocked the path of ability. But under President Hosni Mubarak, the rise of essential talent has been encouraged, and staff training and experience have been reinforced. Thus the elite of the Platonic pyramid has become capable of facing the Israeli air force on equal terms.

Another serious problem, and harder to cure, is the lower-ranks situation. NCOs, the backbone of an air force, have been hard to come by, train, and retain. The necessary yeoman class does not exist, and members of the lower middle class can better themselves in commerce; they may join the EAF only to obtain the training necessary to become civilian entrepreneurs. This problem must be solved by making military service acceptable and desirable, in part by raising the pay to match that of skilled civilians. And the same themes must be extended to the enlisted ranks.

The Syrian air force has not been able to emerge phoenix-like for a variety of internal and external reasons. Although the Iraqi air force has been dismantled, the Iranian and Pakistani have survived more or less intact. The Saudi air force has experienced little combat, and the very small Arab air forces, primarily in Arabia Deserta, have been able to survive because their few needs have been supplied by foreigners while indigenous personnel were undergoing training. Losses have been small and oil revenues large.

A fundamental question is whether Muslim rote learning of the Koran and obedience to the imams are fundamental causes of ineffectiveness. After all, the Muslim religion is more concerned with commercial than with military values. This certainly helps explain the Arab need for detailed plans from the high command and the lack of initiative and flexibility in air combat, as well as the ready adherence to rigid Soviet tactics in the Egyptian and Syrian air forces. But it appears to be more a matter of the level of civilization. Based on the experience in Europe, the United States, China, and Japan, as commerce, education, and technology spread and social strata become porous, societies become more flexible and adaptable. In the nineteenth and early twentieth centuries, the dead hand of the Ottoman elite and the mosque preserved a society that, in the second half of the twentieth century—the period of Israeli technical and military dominance—had to adapt or die. The learning process was terribly expensive, though heavily funded by Soviet and American Cold War largesse. By 2003, the EAF had become a player. The Jordanian air force remains a miniature of its RAF mentor, the Syrian air force is still a marginal factor, the Saudi air force reflects its U.S. tutor, and the other Arabian peninsular air arms remain colonial.

Of all the Arab air powers, the Iraqis most valued the airplane as a symbol of modernization. But those who manned the air force were not trusted, and they had little faith in Saddam Hussein. As a result, the air arm was ineffective against Iran and was overwhelmed and fled in the Second Gulf War; in the Third Gulf War, its remnants were annihilated, and the airplane as a dictatorial symbol was buried.

For Arab air forces to avoid defeat, they have to be seen as heroes in the air supported by a self-respecting, professional Muslim infrastructure that no longer consists of simple Koranized mechanics. Turn that around, and it explains their past defeats.

Suggestions for Further Research

Aside from works on the Israeli air force, one on the Egyptian air force, and an overall view of the Arab armed forces, studies of the Arab air forces are virtually nonexistent, as evidenced by the few works cited below. In part, this is due to language difficulties and to Arab and Muslim secretiveness.

The focus in the Middle East has been on the IAF because of its successes and because of its publicity consciousness. With regard to the Muslim air arms, the researcher can find a fine start in the multivolume, broadly conceived history of the Pakistani air force. There is also a study of the Jordanian armed forces that includes the air force. Otherwise, the field is wide open, and a good deal may be gleaned from international periodicals and perhaps former Soviet sources and Western files as the archives are opened.

There is also the possibility of studying aircraft and weapons procurement in Soviet, French, British, and U.S. sources, including materials in company hands. Such research should include the ancillary industries, electronics, and simulation. Control personnel may also be a source, both orally and in photos and documents. Nor should airfield building be neglected.

There is scope for preliminary studies based on technical and professional journals. Another source is CIA reports and assessments, as well as Western air attaché and other reports being released under the thirty-years rule. Documents pertaining to aircraft and weapons tests of captured aircraft can be found in the U.S. Air Force (USAF) Air Material Command files. The same applies to British, French, and possibly Soviet sources.

The State Department's *Foreign Relations of the United States* should be explored for the relevant countries, as should the holdings of libraries of such institutions as the USAF Air University and the USAF Museum. Other broad sources are the U.S. Army's area handbooks and the U.K. Ministry of Defence MODILS Library Bulletin.

The Russian archives are untouched, and they should have much to reveal about the supply of material and advisers to the Arab air forces and to India.

Any research on Middle Eastern air forces has to focus on politics—of both the region and outside suppliers—and on the milieu in which Arab air forces have operated.

Recommended Reading

Begin with C. P. Snow, *The Two Cultures* (New York: Cambridge University Press, 1964). A solid place to start on the Middle East in general is Deborah J. Gerner and Jillian Schwedter, eds., *Understanding the Contemporary Middle East,* 2nd ed. (Boulder, Colo.: Lynne Rienner, 2004).

The following works also provide general background on the Arab world: Bernard Lewis, *The Multiple Identities of the Middle East* (New York: Schocken Books, 1998), and Raphael Patai, *The Arab Mind,* rev. ed. (New York: Scribner's, 1983).

On the economies, see Roger Owen and Sevket Pamuk, *A History of the Middle East Economies in the Twentieth Century* (Cambridge: Harvard University Press, 1999).

The work edited by J. H. Thompson and R. O. Reischauer, *Modernization of the Arab World* (New York: Van Nostrand, 1966), provides a perspective from the 1960s. Englishman J. C. B. (Sir John) Richmond, a longtime resident of the Middle East, contributes *Egypt, 1798–1952* (New York: Columbia University Press, 1977), and Amos Perlmutter, a sociologist specializing in the military in Israel and the Middle East, gives a provocative but dated view in *Egypt: The Praetorian State* (New Brunswick, N.J.: Transaction Books, 1972). See also United Nations Development Fund, *The Arab Human Development Report* (New York: UN Development Programme, Regional Bureau for Arab States, 2002–2003).

On the recent modernization of Egypt, Timothy Mitchell has exploited the archives for *Rule of Experts: Egypt, Techno-politics, Modernity* (Berkeley: University of California Press, 2002). See also Anwar Sadat, *In Search of Identity: An Autobiography* (New York: Harper and Row, 1978). Glubb Pasha and P. J. Vatikiotis offer *Nasser and His Generation* (New York: St. Martin's Press, 1978), and *The Egyptian Army in Politics: Pattern for a New Nation* (Bloomington: Indiana University Press, 1961). Robin Higham, "Ending Enmities," *War & Society* 11, no. 1 (May 1993): 179–93, focuses on the long term; by the same author, "Air Operations as Guerrilla Warfare," *Defence Analysis* 15, no. 2 (1999): 215–22, provides a modern slant.

There is little English-language aviation material, so interested readers must search for periodical articles in *Flight International, Interavia,* and technical journals or study Arab politics and character on the road to a grand strategy. A recent overall assessment is Anthony H. Cordesman, *Military Balance in the Middle East–vi–Arab–Israeli Balance—Overview* (Washington, D.C.: Greenwood Publishing Group,1998). A start on the Arab–Israeli wars can be made in Ronald M. Devore, *The Arab–Israeli Conflict: A Historical, Political, and Military Bibliography* (Santa Barbara, Calif.: ABC-CLIO, 1976). Also see Edgar O'Ballance, *Battles in a Hostile Land: 1839 to the Present* (London: Brassey's, 2002); John Westwood, *The History of the Middle East Wars* (North Dighton, Mass.: World Publications Group, 1984); and Michael B. Oren, *Six Days of War: June 1967 and the Making of the Modern Middle East* (New York: Oxford University Press, 2002). Major General Chaim Herzog, author of *The Arab–Israeli Wars: War and Peace in the Middle East from the Idea of Independence through Lebanon* (New York: Random House, 1982), saw the struggle as a participant and later as a commentator. A different slant is supplied by Jon D. Glassman, a career foreign service officer who was in Moscow, in *Arms for the Arabs: The Soviet Union and War in the Middle East* (Baltimore: Johns Hopkins University Press, 1975).

The best place to start for the EAF is Kenneth M. Pollack, *Arabs at War: Military Effectiveness, 1948–1991* (Lincoln: University of Nebraska Press, 2002), followed by Ian O. Nordeen and David Nicolle, *Phoenix over the Nile: A History of the Egyptian Air Force* (Washington, D.C.: Smithsonian, 1996). For equipment details in context, see Charles Stafrace, *Arab Air Forces* (Carrollton, Tex.: Squadron/Signal, 1994), and Victor Flintham, *Air Wars and Aircraft . . . 1945 to the Present* (New York: Facts on File, 1990).

On the Royal Jordanian Air Force, see Brigadier S. A. El-Edross, *The Hashamite Arab Army, 1908–1979* (Amman: Jordan Publishing Committee, 1980). There are a few studies of the Jordanian military, notably by the former commander of the Arab Legion, Glubb Pasha, and P. J. Vatikiotis, *Politics and the Military in Jordan: A Study of the Arab Legion 1921–1957* (New York: Praeger, 1967).

On the Pakistani air force, see Anonymous, *The Story of the Pakistan Air Force: A Saga of Courage and Honour* (Islamabad, Pakistan: Shaheen

Foundation, 1988). This is the first of two volumes and covers 1947–1998, including details about both operations and infrastructure. Volume 2 is Shaheen Foundation, *The Story of the Pakistan Air Force, 1988–1998* (Oxford: Oxford University Press, 2000).

The struggles in Lebanon are depicted in Itamor Rabinovitch of Tel Aviv University, *The War for Lebanon, 1970–1985* (Ithaca, N.Y.: Cornell University Press, 1985).

The 1967 war is succinctly discussed in Edgar O'Ballance, *The Third Arab–Israeli War* (Hamden, Conn.: Archon Books, 1971). The fighting after 1967 is covered in Yaacov Bar-Siman-Tov, *The Israeli–Egyptian War of Attrition, 1969–1970: A Case Study of Limited Local War* (New York: Columbia University Press, 1980); Edgar O'Ballance, *The Electronic War in the Middle East, 1968–70* (Hamden, Conn.: Archon Books, 1970); Frank Aker, *October 1973: The Arab–Israeli War* (Hamden, Conn.: Archon Books, 1985); Badvi, Magdoub, and Zohdy (three faculty members at the Nasser postgraduate military academy), *The Ramadan War, 1973* (New York: T. N. Dupuy Associates, 1978); authors guided by Naseer H. Arvri, a professor in America, in *Middle East Crucible: Studies on the Arab–Israeli War of October 1973* (Wilmette, Ill.: Medina University Press International, 1975); and Abraham Rabinovich, *The Yom Kippur War: The Epic Encounter that Transformed the Middle East* (New York: Schocken, 2004).

On the Gulf wars, the following may be consulted: *Aviation Week, The Persian Gulf War: Assessing the Victory* (New York: AW&ST, 1991); Alberto Bin, Richard Hill, and Archer Jones, *Desert Storm: A Forgotten War* (Westport, Conn.: Greenwood Press, 1998); Anthony Cordesman, CSIS Europe Program Congressional Staff Forum Summary: "The Iraq War and Regional Stability" (congressional staff briefing by Anthony Cordesman, CSIS Arleigh Burke Chair in Strategy, March 3, 2003); Richard Ben Cramer, *How Israel Lost* (New York: Simon and Schuster, 2004); Bert Kinzey, *The Fury of Desert Storm: The Air Campaign* (New York: McGraw-Hill for TAB Books, 1991); Edward J. Marolda and Robert J. Schneller Jr., *Shield and Sword: The United States Navy and the Persian Gulf War* (Annapolis, Md.: Naval Institute Press for the Naval

Historical Center, 1998); the staff of *US News, Triumph without Victory: The Unreported History of the Persian Gulf War* (New York: Random House, 1992); and USAF Analysis Division, CENTAF, "Operation Iraqi Freedom—by the Numbers" (30 April 2003).

An Egyptian editor of distinction, Mohamed Heikal, provides a useful corrective to U.S. views of Manifest Destiny in *Illusion of Triumph: An Arab View of the Gulf War* (London: HarperCollins, 1992).

For the view from the Soviet side, see Mark A. O'Neill, "Air Combat on the Periphery: The Soviet Air Force in Action during the Cold War, 1945–1989," in *Russian Aviation and Air Power in the Twentieth Century,* ed. Robin Higham, John T. Greenwood, and Von Hardesty (London: Cass, 1998), 208–35.

Also see Anthony H. Cordesman and Abraham R. Wagener, *The Lessons of Modern War: I. The Arab–Israeli Conflicts, 1973–1989* (Boulder, Colo.: Westview Press, 1990); William J. Olsen, "Air Power in Low-Intensity Conflict in the Middle East," *Air University Review,* March–April 1986, 2–21; and David Segal, "Air War in the Persian Gulf," *Air University Review,* March–April 1986, 50–61.

CHAPTER FOUR

Defeat of the German and Austro-Hungarian Air Forces in the Great War, 1909–1918

John H. Morrow Jr.

The First World War witnessed the rise of airpower and the clash of embryonic air forces in the first struggle for aerial supremacy in the skies above battlefields and even over enemy homelands. Before the earliest days of the conflict, military establishments and pundits envisioned both the tactical and the strategic employment of aerial vehicles.

In the historical analysis of the defeat of the German and Austro-Hungarian air services at the hands of the Entente powers in the Great War, the historian must not allow twenty-twenty hindsight to occasion unduly harsh judgments or to deny what appeared to be viable options at the time. The advantages and disadvantages—the latter perhaps better termed flaws—that were not necessarily apparent to the participants are seen more clearly by the historical observer. Thus, the historian is obliged to point out how those engaged in aviation at the time perceived such circumstances and to show how what might have appeared advantageous or even necessary then now looms as a flaw, perhaps even a fatal one, in retrospect.

Furthermore, the subject requires the historian not to begin with the aerial struggle waged during the war but to return to the origins of the two key intertwined elements in the evolution of air arms—the military aviation establishment and the aviation industry—within the context of their societies. The origins of military aviation prior to 1914 would greatly affect the wartime evolution of military aviation in Germany and Austria-Hungary. Consequently, an analysis of the defeat of these two powers must begin with the prewar conditions that established the context for wartime success or failure. Finally, such an analysis cannot occur in a vacuum but must consider the capabilities of their opponents. Few things are absolute; most are relative in the realm of warfare, as in life.

The Military and Powered Flight to 1914

In Germany the appearance of Swabian Count Ferdinand von Zeppelin's giant rigid airship, or dirigible, during the first decade of the twentieth century colored the evolution of the military in the German empire. These early behemoths, which dwarfed other airships at a length of more than 400 feet, seized the popular imagination by successful flights of twelve hours' duration. When the zeppelin LZ4 crashed attempting a twenty-four-hour flight, the ensuing popular campaign raised more than 7 million marks for zeppelin construction. The enthusiasm of the General Staff, civilian agencies, and Kaiser Wilhelm II—who had encouraged Zeppelin and proclaimed the count "the greatest German of the twentieth century" before its first decade had elapsed—forced the more abstemious and dubious Prussian War Ministry to accept a zeppelin as army airship Z1 in 1909, before it met performance stipulations. It also ensured that the shadow of the cigar-shaped monster would loom over the development of the German army and navy air services through the end of the First World War.

At the end of 1908, when the military effectiveness of the dirigible remained open to question, the airplane burst on the stage of European aviation. Captain Hermann von der Lieth-Thomsen of the German General Staff advised his superiors in 1907 of the danger of merely observing foreign aerial progress, but his negotiations with the Wright brothers for their flying machines foundered on the Prussian War Ministry's continued objection to price. By the fall of 1908, the flights of the Wrights and French aviators made it abundantly clear that the

United States and, more critical for European military developments, France led the way in heavier-than-air flight.

Both the General Staff and the War Ministry agreed on the potential usefulness of aerial vehicles for reconnaissance and communications. Whereas the German General Staff desired to forge ahead with both airplanes and airships, the War Ministry—rather ironically, in view of its reservations about the zeppelin—believed that the giant dirigible would enable the German army to wait until private initiatives perfected a military aerial vehicle. By the fall of 1908, some ten enterprises in Germany were building airplanes. In Austria-Hungary, as in Germany, the General Staff wanted to take the initiative in aviation, while the War Ministry exercised restraint because of the cost and the embryonic nature of aviation technology.

The public's interest in aviation, exemplified by the zeppelin campaign, led to the founding of a German Air Fleet League, modeled after the German Navy League, to promote military aviation in the summer of 1908. By 1909, its membership of 3,000 individuals, including a board of prominent military, industrial, and political figures, demonstrated that military aviation had aroused the public interest. In general, flight was being drawn into the web of increasingly bellicose popular attitudes encouraging the militarization of aviation in the years remaining before the war.

The French army, true to its long history in aeronautics, led other armed forces in its support of powered flight, sponsoring aviation inventors in the early stages of experimentation. France possessed the leading early aero-engine designers and manufacturers, such as Léon Levavasseur and his incredibly light V8 Antoinette engines. Aero engines, the heart of any powered aircraft, determined the boundaries of aerial achievements, and the engine industry's production of more powerful and reliable engines would enable flying machines to become practical instruments of warfare. France claimed the early lead in aviation in the realms of both military attention and industrial achievement.

Yet in 1908, the reliability of aerial machines still lay in the future, and the suitability of embryonic airplanes and airships for military use remained a matter of conjecture rather than fact. Aviation received low military priority, but by the end of 1908, aeronautics appeared on the verge of acceptance by military establishments. The years 1909 to 1914 would witness its transformation into an instrument of modern warfare.

As historian Charles R. Gibbs-Smith observed, "In 1909 powered aviation came of age. The aeroplane became technically mature and established in the public mind . . . , and the beginnings of national aircraft industries, as well as government concern for aviation, were now to be seen throughout Europe." French achievements in 1909—Louis Blériot's crossing of the English Channel in July and the Reims aviation week in August—demonstrated that the young French aircraft and aero-engine industry had gained ascendancy over its rivals in other countries.

The French army began to purchase airplanes in 1909 and employed them in military maneuvers in September 1910. The army also bought three dirigibles in 1909 because of their superior range and load, but high winds and deep valleys in Picardy prevented their flight and landing at maneuvers. The army ordered more than 200 planes in 1910 and 1911—Farmans, Blériots, and Voisins—powered by Renault V-8, Gnome seven- and nine-cylinder rotary, and Anzani three-cylinder fan-type engines. The bedrock of French aviation supremacy through 1914 would lie in its vaunted aviation industry—in aircraft firms such as Farman, Voisin, Caudron, and Nieuport and, in particular, in its aero-engine industry of Gnome-Rhône, Renault, and Salmson, which produced rotary, V-8, and radial engines, respectively.

Between 1909 and 1911, the German army made crucial choices about dirigibles and airplanes. Both commercial and military airships, the latter of which still exhibited inadequate reliability and performance and excessive dependence on weather, crashed in 1910 and 1911. The German aviation companies tended to copy foreign aircraft designs, neglected engine development, and relied on French and Austrian Daimler engines until German Daimler began to produce its Austrian subsidiary's engine in 1911.

The German General Staff sought to form an aviation agency and purchase the best airplanes, because Lieth-Thomsen believed that heavier-than-air craft would soon supplant airships. The War Ministry, however, insisted on developing both large and small airships as well as airplanes because of the army's sizable investment in contracts, trained personnel, and sheds for airships. The ministry reasoned that French supremacy in airplanes rendered the preservation of German supremacy in airships all the more essential. German ruling circles, which regarded the airship as a symbol of German aerial superiority and a means of pressure and threat against foreign countries, were reluctant to admit failure. Airship firms maintained excellent contact with the War Ministry through their employment of former officers. The airship manufacturers also threatened to sell airships to Germany's potential enemies, Britain and Russia, both of which lagged behind in the aerial arms race.

The War Ministry's policy of dividing its resources led to an impasse with the airplane industry: military investment remained contingent on the industry's prior development of airplanes, but the industry required the state's money to develop such planes. The German Air Fleet League, which had grown to nearly 12,000 members by 1912, and the German Aviators Association, which numbered 60,000 members, lobbied for the aviation industry, but German military contracts for airplanes remained meager.

In Austria-Hungary, the General Staff and the War Ministry assumed the same respective roles of airplane promoter and restrainer as their German counterparts. Austria-Hungary had no airship firms, but it possessed designer Igo Etrich's Taube (dove) monoplane and Austro-Daimler designer Ferdinand Porsche's six-cylinder in-line engine, both of which the German and Austro-Hungarian armies would adopt in 1911. Although civilian aviation interest groups and the General Staff pressed for the military adoption of the airplane, such sentiment did not percolate through the ranks. The colonel commanding aspiring Austrian military aviator Lieutenant Emmanuel Quoika inquired pointedly, "Do you want to become an aerial acrobat or an artillery captain?" and ordered Quoika to sell the airplane he had recently purchased.

Between 1909 and 1911, Germany continued to rely primarily on the dirigible, although Austro-Daimler did give Germany and Austria-

Hungary a more fuel-efficient, if heavier, engine than its French counterparts. The zeppelin's range, which rendered it better suited for strategic reconnaissance than the airplane, justified the military's support, yet the War Ministry's refusal to stop funding smaller airships in favor of airplanes lessened the potential funds for the latter while doing nothing to impede a Zeppelin monopoly of giant airship production.

The steady expansion of the German army air arm and its aviation industry continued from 1912 through 1914. The army, as well as the German navy, divided resources for aviation between the rigid airship and the airplane, as both the General Staff and the Admiralty Staff had high expectations of their zeppelins' performance. By April 1912, the German army had ten airships, and in September the General Staff planned to have fifteen ships for bombing by 1914, although it had yet to test their effectiveness as bombers. Count Zeppelin's exaggerated claims and, more crucially, the General Staff's own determination to preserve its lead over France in airship development shaped the army's plans to use airships as bombers more prominently than did hard evidence.

By 1912, German General Staff Chief Helmuth von Moltke was determined to have as many airships as possible operational for a future war because of his exaggerated notions of the zeppelins' "first-strike capability." On 24 December 1912 he informed the War Ministry: "In the newest Z-ships we possess a weapon that is far superior to all similar ones of our opponents and that cannot be imitated in the foreseeable future if we work energetically to perfect it. Its speediest development as a weapon is required to enable us at the beginning of a war to strike a first and telling blow whose practical and moral effect could be quite extraordinary."

Aviation journals echoed such sentiments. For example, articles in the *Deutscher Luftfahrer Zeitschrift* in 1913 anticipated pinpoint and unstoppable zeppelin attacks on enemy targets in the dark of night. German war plans in 1913 placed zeppelins directly under the control of the high command and the army commands for strategic reconnaissance and bombing missions. Airships participated in military maneuvers and tested the installation of machine guns and wireless sets but performed only one bombing trial before the war.

In 1913 and 1914 zeppelins met the army's minimum performance standards only under the most favorable of weather conditions, and an army ballistics expert warned that dirigibles would be vulnerable to incendiary shells. In the summer of 1914 the army had only seven airships suitable for military operations. This inability to adhere to the General Staff's optimistic plan should come as no surprise. The monstrous zeppelins were expensive, difficult to build, and awkward to house. The Z9 delivered in July 1914 was 158 meters long, 22,740 cubic meters in volume, and powered by three 200-horsepower engines. It cost 860,000 marks, the equivalent of thirty-four biplanes.

The General Staff considered airplanes suitable for shorter-range reconnaissance, communications, and artillery spotting, and Moltke pressed for the maximum number of airplanes possible. The War Ministry increased airplane contracts from 130 in 1912 to 432 in 1914, as the number of aircraft factories increased and became larger in scale and financially sounder. Although the General Staff proposed the formation of an inspectorate of military aviation separate from transport, the War Ministry rejected such a radical measure. Consequently, as of 1914, the aviation inspectorate remained under the control of the transportation inspectorate.

In 1914 the army's aircraft firms, in order of productivity, were Albatros, Aviatik, Luftverkehrsgesellschaft (LVG), Euler, Gothaer Waggonfabrik (Gotha), Jeannin, Fokker, and Luftfahrzeuggesellschaft (LFG). These firms, with the exceptions of Euler and Jeannin, would form the nucleus of the German wartime aircraft industry. The larger factories, in which mechanization and serial production were common, employed a few hundred workers. The inspectorate had also standardized monoplane production on the Taube, thereby ossifying its design at most manufacturers, with the exception of Anthony Fokker, a maverick young Dutchman who built primitive light monoplanes from 1912 to 1914.

All these firms relied on one aero-engine company, Daimler, which produced one engine type—the 100-horsepower, six-cylinder, water-cooled in-line. The engine was excellent—sturdy, reliable, and fuel efficient—but the monopoly was unhealthy. In a belated effort to lessen dependence on Daimler, the War Ministry included the Benz company in its engine orders in June 1914. The aviation inspectorate, which lacked aero-engine experts, had stipulated a maximum of 125 horsepower for army aircraft, thus actually discouraging the industry's development of more powerful engines for airplanes. The army also belatedly showed interest in the Oberursel company's 80-horsepower rotary engine, a copy of the French Gnome.

The German army's effective control of civilian aviation through its pervasive influence in German society and its ability to play on chauvinistic notions of cultural supremacy to bolster military aviation had enabled it to militarize prewar German aeronautics. A National Aviation Fund of 1912 raised more than 7.2 million marks, which bought airplanes for the army; financed the training of military aviators; built airfields destined for military control in wartime; provided partial funding for a civilian German Research Institute for Aviation, a central research agency that performed stress tests on military aircraft; and sponsored "civilian" flight competitions that became virtually long-range reconnaissance flights for slow, stolid military two-seat biplanes and monoplanes and their crews. Civilian sport aviation associations zealously defended the army's priorities and effectively thwarted the efforts of a few airplane manufacturers to use these contests to promote sport aviation and a market for light, fast airplanes.

In August the German army went to war with ten dirigibles and 245 rugged, dependable two-seat reconnaissance aircraft with reliable water-cooled in-line engines. Its training techniques, using civilian flight schools and flight competitions to conduct preliminary training and constant tests for its military pilots, had resulted in 254 pilots and 271 observers. The Germans had actually overtaken the French airplane force through their single-minded development of reconnaissance planes.

The German naval air arm remained in an embryonic state in July 1914, although it now had two firms—Flugzeugbau Friedrichshafen and Hansa-Brandenburg, the latter directed by Ernst Heinkel—to build floatplanes suitable for operations in the rough waters of the North Sea. It had also successfully encouraged the development of Mercedes 150-

horsepower engines for its heavier planes. The Admiralty Staff regarded airplanes as useful for reconnaissance and defense against enemy aircraft, and it considered zeppelins cheaper substitutes for cruisers to perform reconnaissance for the fleet and as potential bombers against enemy coastal installations. Yet the navy had only one airship with which to conduct bombing trials, after its first two crashed and burned.

In Austria-Hungary grandiose designs continued to contrast with meager state finances and industrial development from 1912 through 1914. The army's sole plane, the Lohner Pfeilflieger (Arrowflyer), which held the world altitude record for a time, had to be grounded in March 1914 because a series of crashes indicated that its frame was too fragile to meet the army's stress requirements. German aircraft firms rushed to fill the production vacuum, but too late to become a factor in mobilization. Consequently, the Austro-Hungarian army had only forty planes and eighty-five trained pilots at the outbreak of war. At least Lohner had developed a viable flying boat suitable for the waters of the Adriatic Sea, and the Austro-Hungarian navy had some twenty-five pilots and ten to fifteen usable seaplanes in August 1914.

At the outbreak of war, the disjunction between the fantastic goals set for zeppelins and the paucity and inadequacies of the airships, not to mention the failure to develop bombs or bombing techniques, was common to both the German army and navy. The army viewed its few airships as wonder weapons, but the hope that its minuscule zeppelin fleet could deliver a telling first strike revealed a great gap between expectations and reality. The army's development of increasingly larger airships stemmed from technological and military considerations, as the ship's size determined its load, range, and ceiling.

The military bureaucracies of the various combatant powers did not function monolithically or systematically in their adoption of aviation: in Germany and Austria-Hungary, general staffs and war ministries differed over aviation; in France, engineers and artillerymen fought over control of the new arm; and in Britain, the senior service—the navy—contested the army's claim of control. As the air arms grew in importance and size, issues regarding their autonomy arose, even within the army. In Germany, aviation remained subordinate to transport, while in France, the engineer-artillery struggle continued even after aviation achieved an allegedly autonomous Twelfth Directorate within the French War Ministry.

The origins of military aviation indicate the importance of linking doctrine and aircrew training with technological and industrial progress. By 1914, all the powers had trained cadres of pilots, but although they had all decided to employ the airplane for reconnaissance, only Germany and France had been training observers since 1911. Apparently, none of the armies deemed it necessary to train army commanders to appreciate and understand the employment of aviation. On the whole, a limited doctrine restricting the airplane's use to reconnaissance and communications accorded well with the technological and industrial realities of the moment. On the one hand, doctrine could lead to dead ends and wasted effort if it exceeded the state of technology or to inflated expectations if actual maneuvers did not test it—as in the zeppelin's case. On the other hand, the German air arm's development of a standard reconnaissance airplane linked early doctrine too closely to materiel procurement, as it focused the industry's efforts too narrowly on a single aircraft type—the slow, stable two-seater.

Economic as well as military doctrine determined the armies' relationship to the aviation industry. The military attempted to control aviation production, most fundamentally through its position as the prominent or sole consumer in the prewar market. All armies professed a preference for relying on a private aviation industry. In France, the army dispersed its contracts among its major aircraft firms; in Germany, the military required more companies and drew large industrial combines such as AEG and Gothaer Waggonfabrik to aircraft production. Both limited their industries to a few proven companies.

Timing partially determined the above relationships: the French airplane firms antedated military interest in aviation, as did the zeppelin in Germany; elsewhere, the aviation industry formed in response to military interest in flight. Consequently, the French army was preoccupied with controlling the industry, and the German army was focused on its control and expansion. The British army, with little industry at all, relied on the Royal Aircraft Factory in the relative absence of private companies, but the Royal Navy used private contractors.

The French airplane industry, with some 3,000 workers in nine firms, was larger than both Germany's eleven firms employing 2,500 workers and the British industry comprising twelve firms and only 1,000 workers. The French army eschewed standardization and obtained variety of type at the expense of uniformity of production, supply, and repairs. The German army insisted that firms produce standard monoplane or biplane types but allowed diversity of construction, thus achieving neither uniformity of production nor variety of type. The British, far behind both, would need French assistance in aircraft and particularly aero-engine supply. Among the lesser European powers, Russia and Italy possessed individual designers of merit but were dependent on France, while Austria-Hungary relied on Germany. The United States, land of such aerial pioneers as the Wright brothers and Glenn Curtiss, had the smallest aircraft industry of all.

Future industrial expansion in wartime would hinge on a reservoir of highly skilled craftsmen. In the production of aviation engines, France could turn to its vaunted automobile industry—which was a distant second to the United States in prewar output—because automobile firms such as Renault and De Dion Bouton were already producing aviation engines. The Daimler monopoly rendered German aero-engine supply more precarious, and the German automobile industry ranked third in the world behind France. Furthermore, although the efficiency and long service life of German six-cylinder, liquid-cooled, in-line engines made them ideally suited for states that would experience wartime shortages of labor, material, and fuel, they did not offer the diversity that French in-line, rotary, and radial engines did.

France and Germany were clearly ahead of other nations in military aviation, but historians do could not agree on which one led. French aviation historians, for example, credited the Germans with wresting qualitative leadership from the French, whereas German historians proclaimed France ahead. Some British authors asserted that Sopwith's victory in the

Schneider Trophy Race in April 1914 demonstrated that the lead in aviation was gravitating to Britain, but these claims were premature.

The German army went to war with some 250 of 450 airplanes considered frontline materiel; French frontline squadrons contained only 141 of some 600 military airplanes. Yet in 1913 the French had expanded their general reserve, which comprised frontline airplanes, to nearly the same size as the frontline squadrons. The air services were thus nearly equal in size, and if the Germans had marginally more active-duty field pilots—254 to 220—the French probably had more reservists available, given their earlier start in aviation and larger number of civilian pilots. The size of the air service represented only one factor in preparedness in a service based on technology and industry. The Germans had the zeppelin, but the French led in aero engines, and without resorting to hindsight (which would show the engine to be far more crucial than the airship for wartime aviation), one can state that France and Germany, the two powers that were the nuclei of their respective alliances' aerial strength in 1914, were about equal.

A comparison of their supporting casts, however, makes it evident that France, allied to Britain and Russia, was in a far stronger position than Germany, which was tied to Austria-Hungary. Austria-Hungary possessed the weakest air service and aviation industry of the combatant powers. Austria possessed individual designers of genius, such as Ferdinand Porsche, but it lacked the industrial prowess to manufacture materiel in quantity.

The military concentration on reconnaissance and artillery spotting led armies to prefer slow, stable, two-seat airplanes, primarily biplanes because of their superior sturdiness in an era of externally braced wings. Their top speed in 1914 was 65 to 75 miles per hour, far below the 1913 speed record of 126.67 miles per hour set by a French Deperdussin monoplane. Armies clearly distinguished between sport and military aviation in airplane type and pilot training, and the untrammeled individualism and daredevilry associated with sport aviation were anathema to the military's emphasis on the disciplined fulfillment of one's mission as part of a unit.

Prewar military aviation agencies had sought to minimize flight risks through the standardization of aircraft controls, the installation of instruments, the supervision of construction, and the introduction of safety standards and stress tests. Aviation science institutes—the German Aviation Institute and Kaiser Wilhelm Institute in Germany, the Eiffel Institute and Meudon in France, and the National Physical Laboratory and Royal Aircraft Factory in England—were all connected to or part of the military establishment and advised military procurement agencies on such matters as break tests for airplanes.

Ultimately, the armies bought slow, stable planes for average pilots; these were not the great and gifted sport pilots whom the public worshipped as heroes and whom some army officers regarded as acrobats. Ironically, the war would catapult the flying machine into the forefront of the public imagination as the vehicle of the aces, who were reminiscent of the prewar aviators who had mastered the skies. Wartime aerial combat, the realm of aviation in which the least prewar experimentation had occurred, would reintroduce aspects of sport aviation that the military had tried to escape—the individual exploits of gifted pilots and the high-performance airplane, which was faster, more maneuverable, riskier to fly, and occasionally even dangerous to its pilots. The aura of prewar sport would return, but in a context so deadly that it altered the nature of the game.

The Great War, 1914–1918

From the start of the conflict at the beginning of August, German airplanes acquitted themselves well on the Western and Eastern Fronts. In the west they kept track of the French retreat and played key roles in detecting enemy troop movements. Aviation was even more important on the Eastern Front because, in light of the great numerical superiority of the Russian cavalry, airplanes provided key information on Russian troop movements in the relative absence of cavalry units. German aviators were instrumental in preparing the victory at Tannenberg and in preventing a more serious defeat on the Marne. By the end of August, the airplane had developed from "a supplementary means of information relied upon principally for confirmation" to "the principal means of operational reconnaissance—an important factor in forming army commanders' decisions."

German army airships, in contrast, proved vulnerable to ground fire during reconnaissance and bombing attacks on the Western Front. The zeppelins' initial wartime performance was disastrous: five of the seven military airships were destroyed by early October 1914, four by fire from the ground. The giant dirigibles had become liabilities; they could fly over the front only on dark nights. In addition, replacement of their losses was slow, with only three new zeppelins entering service by the end of 1914. Though the airship failed the German army, the German naval airship officers considered it a cheap substitute for cruisers to scout for the fleet. Obsessed with the idea of bombing England, the naval leadership expanded its tiny unit, which had only one airship at mobilization.

Organizational deficiencies lessened the potential efficacy of German aviation units. The high command and army commands lacked central aviation agencies to collect and assess fliers' reports; the subordination of aviation to transport deprived aviators of direct access to the front command; and the practice of equipping units with different aircraft types complicated supply. In the rear, the existence of war ministries in some of the smaller German states, Bavaria in particular, also impeded the most efficient mobilization of the aviation industry. More crucially, the Prussian War Ministry, although it insisted on increased aircraft deliveries in early August, impeded the industry's ability to comply by resisting aircraft price increases until February 1915, four months longer than did the French War Ministry, which had issued extensive aircraft contracts and consented to price increases in October 1914.

Whatever the problems in Germany, they paled before the inadequacy of Austro-Hungarian aviation, which quickly became a burdensome appendage to its German counterpart. The tiny size of the Austrian aircraft industry forced the Austro-Hungarian army to resort to the German industry for airplanes, but the army refused to extend advance payments to manufacturers to allow them to procure materials. The army also demanded the immediate development of a large twin-engine biplane (the Grossflugzeug, or G-plane) from an industry that was barely able to deliver ten airplanes a month.

The onset of the war disrupted aviation production everywhere, and the French responded most quickly to the new challenges. In October the French army began specializing its aircraft types as it geared up aircraft and particularly aero-engine production for the anticipated campaign of 1915.

In that year, as the war spread and the consumption of materiel exceeded expectations, resulting in production and labor shortages, the combatant powers confronted critical decisions regarding the mobilization of aviation technology and industry that would have far-reaching implications for the future course of the air war.

At the recommendation of the German high command, the Prussian War Ministry established a chief of field aviation *(Feldflugchef)* on 26 April 1915 to direct all aviation, including the "systematic mobilization of the aviation industry." French attacks in 1915 forced the German army and its air service on the defensive, and the French remained superior in numbers throughout the year. The appearance in the late spring of the C-plane, the standard two-seat biplane equipped with 150-horespower engines and a machine gun for the observer, and then of the single-seat Fokker monoplane *(Eindecker),* armed with a synchronized forward-firing machine gun, kept the German air arm qualitatively abreast of its opponents. The performance of German aircrews, particularly pilots such as Max Immelmann and Oswald Boelcke, demonstrated the quality of their training, and outstanding fighter pilots began to appear on both sides of the Western Front.

The army, confronting continued zeppelin losses, transferred most of its airships to the vast reaches of the Eastern and Balkan Fronts. The navy, which secured the kaiser's consent to bomb England, exaggerated the significance of its initial airship raids in January and was determined to procure increasingly larger zeppelins. An editorial in the *Kölnische Zeitung* on 21 January exulted: "The most modern air weapon, a triumph of German inventiveness . . . , has shown itself capable of carrying the war to the soil of old England! . . . This is the best way to shorten the war, and thereby in the end the most humane." Meanwhile, during the year, the zeppelins grew larger, expanding from 1 million cubic feet in volume powered by four engines to nearly 2 million cubic feet with six engines. These ships could outclimb opposing airplanes, but they remained vulnerable to weather in the air and fire in their sheds on the ground. With the onset of winter, the naval raids ceased, and the airships were left to face the ravages of high humidity, mold, and putrefaction in their hangars.

In perhaps the most significant development for the future of aviation science and technology, Hugo Junkers spent 1915 developing the first all-metal airplane. Its metal construction would make it more resistant to weather and fire, and its thick airfoil offered little more drag and better stability than the thin airfoils used by other airplane designers. The War Ministry tendered Junkers a contract in May, but wartime conditions made it difficult for the firm to find sufficient skilled workers, so the prototype was not finished until December.

Spurred by military contracts, the German aircraft industry increased production from 1,348 airplanes in 1914 to 4,532 in 1915, while engine factories delivered 5,037 engines. Yet German aero-engine manufacture stagnated during the first three quarters of 1915. Erroneous decisions by the inspectorate's engine section, whose one engineer, Walter Simon, managed relations with the industry in a "peculiar and arbitrary manner," compounded the limitation posed by Daimler's near monopoly of production. On 16 November 1914 the inspectorate had decided against the development of engines greater than 150 horsepower to avoid disturbing production. This decision, though prompted by limited production capacity, delayed the evolution of more powerful engines.

In 1915 the inspectorate declined a Benz eight-cylinder, 240-horsepower engine because it did not conform to the six-cylinder standard, but it then allowed Daimler to produce an eight-cylinder, 220-horsepower, in-line engine following its bench test in December 1915. After only forty-two of the engines had been made, however, it halted production because the engine's length caused a lack of stiffness in its crankshaft. In general, shortages of skilled labor and machine tools kept crankshaft manufacture insufficient to increase overall engine production, so the entire year was spent expanding factory capacity. Such inadequate production condemned the German air service to, at best, a temporary aerial superiority over a circumscribed area of the front.

The Austro-Hungarian army's air service encountered even greater difficulty than the Germans. After the Italian declaration of war in May 1915, it was involved on three fronts—Russian, Balkan, and the mountainous Italian. Although the navy's Lohner flying boats dominated the air over the Adriatic Sea in 1915, the limited capacity of the domestic aviation industry kept its monthly deliveries to the army air service below fifty planes. Only the intervention of Hansa-Brandenburg, an Austrian-owned German firm with a superior designer, Ernst Heinkel, saved the army air service. The German army released the Hansa-Brandenburg factory for German naval and Austro-Hungarian production, and the Austro-Hungarian army quickly became dependent on the firm for fast, modern planes. The owner of Hansa-Brandenburg, Camillo Castiglioni, also owned aircraft firms in Vienna and Budapest. Thus, the prospect of monopoly loomed, and Hansa-Brandenburg charged comparatively high prices to the Austro-Hungarian army, whose desperate straits allowed no alternative.

The circumstances in Germany and Austria-Hungary contrasted to those in the Entente powers. French aircraft manufacturers delivered 4,489 planes in 1915—a figure comparable to their German counterparts; French engine production, which rose steadily in the fall and winter, totaled 7,096 in 1915. By 1916, the French were producing nearly two engines to every one airplane, a ratio they sustained for the rest of the war and that the other combatants found impossible to duplicate. Furthermore, these engines were rotaries, radials, and in-lines, a variety unequaled by other countries. French aero-engine manufacturers had established their ascendancy in design and production.

The French aviation procurement agencies also capitalized on innovative designs in 1915. C. Martinot-Lagarde, chief of the engine service, pressed for the development of 200-horsepower engines of V or radial design, because their shorter crankcases and crankshafts in comparison with in-line engines would result in lighter and stronger power plants. He also pressed French firms to evolve materials and accessories such as special steels, aluminum, magnetos, and spark plugs that France, like the rest of Europe, had imported from Germany before the war.

French procurement experts traveled to Spain to purchase a new Hispano-Suiza 150-horsepower V-8 engine, a revolutionary new liquid-cooled engine design. Powerful, rigid, light, and durable, the Hispano-Suiza would deliver ever higher horsepower and would become one of the war's greatest fighter engines.

French aero-engine experts and manufacturers in 1915 thus laid the foundation for the future superiority of French aviation in the First World War. Although Britain still lagged in the development of its aviation industry and remained dependent on France for aero engines, it possessed the industrial resources for substantial expansion. The automobile firm Rolls-Royce designed and began to deliver two high-horsepower engines in 1915—the 250-horsepower Eagle and the 200-horsepower Falcon. Finally, the Royal Flying Corps (RFC) and Royal Naval Air Service were developing a group of accomplished and aggressive aviators. As for the Austro-Hungarian air service, even the Russian aerial effort remained superior to it. Italy's entry into the war in May 1915 brought an embryonic airpower with an air theorist and an aircraft designer—Giulio Douhet and Gianni Caproni, respectively—who focused on the use and development of strategic bombers. With every year, the odds against German and Austro-Hungarian aviation rose.

In 1916, the year of the great battles of Verdun and the Somme on the Western Front, the German air arm was hard-pressed to compete against an enemy that was superior in number and often in aircraft quality. France's Nieuport outclassed the German planes, and Britain's pusher DH2 biplanes were comparable to the Fokker monoplane. By spring, the army forbade flights of the expensive zeppelins over the Western Front, because *Feldflugchef* Colonel Hermann von der Lieth-Thomsen believed that their high losses and negligible results pointed inescapably toward the abandonment of airship operations. The navy, however, persisted in its determination to use zeppelins to bomb England, although RFC airplanes had improved sufficiently to be able to shoot down zeppelins in 1916. At least the zeppelins proved to be effective scouts for the High Seas Fleet. German naval seaplane forces were also acquiring capable floatplanes, in particular the Hansa-Brandenburg W12 *(Wasserflugzeug),* whose speed and maneuverability would give the British a rude shock over the Flanders coast in 1917.

Fortunately for German aviators, the Albatros factory had incorporated plywood construction techniques pioneered in 1913 by the Russian designer Steglau to build a strong, stiff, sleek fuselage in which it installed a Mercedes 160-horsepower, six-cylinder, in-line engine and twin forward-firing machine guns synchronized to fire through the propeller. The resulting biplane (and then sesquiplane) fighter restored qualitative supremacy to the German fighter arm from fall 1916 through spring 1917.

On the home front, the War Ministry improved its procurement hierarchy by adding organs for aircraft acceptance at the factories, secured skilled workers through exemptions in the face of a serious labor shortage, and in the fall of 1916 prepared to select its best aircraft types for licensed production. Hugo Junkers confronted numerous problems in perfecting his all-metal airplane in 1916, among them wartime shortages of engineers, skilled workers, materials, and capital.

Yet the most serious industrial problem remained an inadequate aero-engine production, consisting of 7,823 engines in 1916. Although air chief of staff Lieth-Thomsen wanted the industry and inspectorate to develop and test powerful engines of more than six cylinders, the industry proved incapable of copying either the Hispano-Suiza V-8 or the Rolls-Royce V-12 engines. The German industry had perfected six-cylinder in-lines of 150 to 160 horsepower, but German authorities had not attracted additional firms to engine production or pushed the production of higher-horsepower engines, as the French had. The airplane industry remained competitive, building such sophisticated craft as the Albatros and Junkers, and delivered a total of 8,182 airframes, but the German engine industry was unable to keep pace with domestic aircraft production, much less with its French opponent.

The *Feldflugchef* failed to secure a unified independent air arm, although the air service officially became the *Luftstreitkräfte* (air forces) and gained a commanding general (*Kogenluft,* or *Kommandiere General der Luftstreitkräfte*). Yet German prospects looked grim, as a memorandum written by Air Chief of Staff Lieth-Thomsen on 31 August 1916 and released in early October acknowledged. The Allies, supported by the "world raw material market and the American motor and aircraft industry," had increased their "oppressive" numerical aerial superiority, causing severe losses to Germany's best and most experienced fliers. The German air arm would have to improve its organization and equipment merely to assert aerial control at decisive points in future battles. Lieth-Thomsen was resigned to a defensive posture with occasional and limited aerial ascendancy. It was now questionable whether efficient organization and aircraft technology, the hallmarks of German aviation excellence, would suffice to offset the Allies' mounting numerical superiority.

With the defeat of Serbia and the impending collapse of Russia by late 1916, Germany's Austro-Hungarian ally could focus on the war against Italy. The dual monarchy's air services did not lack for intrepid spirit in its pilots, but there were just too few of them. Shortages of skilled labor and raw materials severely impeded domestic production, which rendered Austria-Hungary increasingly dependent on Hansa-Brandenburg and its Austro-Hungarian subsidiaries' production, just as the German navy increasingly required Hansa-Brandenburg to meet its own needs. Under such circumstances, it made little difference that Porsche developed a 250-horsepower, twelve-cylinder V and a 360-horsepower, six-cylinder engine in 1916. The industry, which delivered fewer than 1,000 airplanes in 1916, could not build them.

In France, the administration and politics of aviation were riven with conflict, and although the aircraft industry delivered fewer planes (7,549) than its German counterpart, French engine production totaled 16,875. These numbers included such advanced types as the Hispano-Suiza, a heavier Renault twelve-cylinder engine, and the

Salmson Canton-Unné radial, all of which would be delivering more than 260 horsepower by 1918. These engines would power, respectively, the famed Spad 7 and 13 fighters, the superlative Breguet 14 reconnaissance-bomber, and the durable Salmson 2A2 reconnaissance biplane. Gnome-Rhône and Clerget continued to press the evolution of rotary engines, which the Germans found difficult to copy because of their inability to secure castor oil, the engine's essential lubricant.

Meanwhile, the British pressed forward relentlessly, particularly in their insistence on an offensive policy at the front, regardless of the inadequacies of their aircraft in 1916. The outcome of the air war did not hinge on aircrew quality, because the aircrews of all countries were volunteers. Although the pressures of attrition often meant that they did not receive complete and extensive training, they did not fail to undertake their assigned missions. By 1916, the British aviation industry began to hit its stride, realizing its potential as the world's leading industrial nation before the war. The Rolls-Royce Eagle and Falcon became superlative high-horsepower combat engines, although their complexity inhibited their production in sufficient numbers and rendered the British dependent on the French for aero engines. Furthermore, three aircraft prototypes appeared in late 1916—the SE-5 and Sopwith Camel single-seat fighters and the Bristol F-2 two-seat reconnaissance fighter—that would be the mainstays of the Royal Flying Corps and, in 1918, of the Royal Air Force through the end of the war. Britain produced over 5,000 airplanes and a similar number of engines in 1916 as it began to assume a role in aviation more commensurate with its potential.

Even the Russian aviation industry, which was in decline, stepped up its production of airplanes and engines in 1916. Italy, with its military focusing on the development of aviation and an industry endowed with the Caproni bomber firm and the Fiat engine company, had also outdistanced Austria-Hungary in military aviation by the end of 1916.

As 1916 drew to a close, Germany was clearly in danger of being overwhelmed on the Western Front if the Allies could translate this industrial superiority into aerial mastery. The airplane had now become indispensable to the conduct of hostilities, and the rapidly growing war of attrition in the air, like the one on land, would be determined by the combatants' success in mobilization. The French and British were already pursuing the aerial offensive, while the Germans husbanded their slender resources, fought defensively, and concentrated their aerial forces to seek an occasional mastery limited in time and space over the battlefield.

For example, the Germans focused on the British at the Battle of Arras in April 1917. Outnumbered two to one in fighter strength and three to one in total aircraft strength over the battlefield, German fighter forces, led by ace Manfred von Richthofen, countered by shooting down 151 planes while losing 66 of their own during "Bloody April." The British official history later concluded that German dominance demonstrated the importance of aircraft performance in gaining aerial superiority and that an air offensive would not necessarily ensure local superiority against a determined and skillful enemy that was numerically weaker but better equipped. Unfortunately for the Germans, the War Ministry's inspectorate contented itself with procuring slightly improved versions of the Albatros, which the new British fighters outclassed. The new Fokker triplane, whose exceptional maneuverability made it a deadly weapon in the hands of such superior pilots as Richthofen and Werner Voss, proved susceptible to wing failure and shoddy construction and consequently spent much of its short career grounded.

The Germans also introduced "infantry fliers," ground attack squadrons equipped with highly maneuverable and well-armed two-seat biplanes and armored monoplanes, including the nearly indestructible Junkers J1 "furniture vans" *(Möbelwagen),* which served effectively in offensive and defensive capacities to assist German infantry. Finally, German long-range reconnaissance crews, who flew alone over the lines, received such superior aircraft as the Rumpler C7, whose six-cylinder, 240-horsepower, high-compression engine enabled it to fly at altitudes of 20,000 feet, above Allied interceptors. Although heavy losses in planes and crews in the 1917 war of attrition forced the air arm to send aviators to the front with less training, and shortages were affecting the quality of materials used in airplanes and engines, the best and most experienced crews and their airplanes remained highly effective.

Now the Germans were waging two uncoordinated strategic campaigns against Britain—the army with giant airplanes (the Grossflugzeuge and Riesenflugzeuge), and the navy with its monstrous zeppelins. The twin-engine G-planes, with their 78-foot wingspans; the four-engine R-planes, with their 138-foot wingspans; and the six-engine, 2 million cubic foot zeppelins represented incredible expenditures of money and materials for no decisive success.

The general German mobilization of the Hindenburg Program actually exacerbated Germany's shortage of raw materials and transportation delays, and it did nothing to reduce the barriers among the German states to the most efficient allocation of skilled labor. The War Ministry's inspectorate increased standardization, licensed production, drew automotive firms to engine production, and prioritized the allocation of materials to favor more productive companies. As raw material prices skyrocketed and workers began to strike, the aviation firms resisted rationing, hoarding gasoline, spare parts, and materials by late fall 1917. Meanwhile, the air service's fuel allotment, instead of doubling during the year to 12,000 tons, plummeted to 1,000 tons in November as the war severed Germany from its sources of oil.

Although the German engine industry never moved beyond the six-cylinder in-line and a few rotary engines, most of which copied French designs, at least the high-compression six-cylinder engines of the Maybach firm and of a new firm, BMW (Bayerische Motorenwerke, or Bavarian Motor Works), promised high output at high altitude. The German military-industrial complex in aviation fell far short of the optimistic production goal set by the Hindenburg Program in 1916: 1,000 airplanes a month by spring 1917. Extant documents indicate that the industry, dogged by shortages of material and labor, attained that level of production only once in 1917. Consequently, when military chief Ludendorff proclaimed an Amerika-Programme in June 1917 stipulating monthly production of 2,000 airplanes and 2,500 engines by January 1918, its fulfillment seemed highly unlikely.

Austro-Hungarian aircraft and engine deliveries to the military remained abysmally low in 1917 and did not exceed a total of 1,300 for the entire year. Inadequate worker exemptions and raw material shortages caused these low production figures, but a desperate Austro-Hungarian army high command, quite out of touch with reality, stipulated in the fall that production should rise to 750 planes and 1,000 engines a month in 1918, including giant and armored planes.

French aviation in 1917 experienced a difficult and tumultuous year on the fighting and home fronts. Delays in the introduction of new aircraft and higher-powered engines left its aircrews, particularly reconnaissance and bomber squadrons, equipped largely with obsolete aircraft. At home, constant bureaucratic changes and vitriolic political clashes contrasted negatively with the relative stability of military control in Germany. Amid this bureaucratic instability and labor unrest, the French aviation industry still managed to produce 14,915 airplanes and 23,092 engines in 1917. The military increasingly concentrated production on Spad fighters and their Hispano-Suiza engines and on Breguet 14 reconnaissance-bombers and their Renault power plants, in preparation for the military campaign of 1918. The key to success in both combinations was the perfection of the engines, which would determine the performance of the aircraft.

British fighter pilots had begun 1917 badly against their German opponents, being rushed to the front with inadequate training to replace high losses, but as the year continued, the new fighters improved their ability to execute long-distance patrols over German airspace. The Hispano-Suiza–equipped SE-5, the Sopwith Camel equipped with Clerget and then improved Bentley rotary engines, and the Bristol fighter with the Rolls-Royce Falcon would serve effectively to the end of the war, although the Camel was increasingly relegated to low-altitude ground attack operations. Bomber and reconnaissance crews, like their French counterparts, made do with inferior aircraft that offered easy prey to German fighters.

At home, the Royal Flying Corps found itself the subject of an increasing determination to combine it with the Royal Naval Air Service and create a separate air force, but the key to the air arm's future performance in the war lay primarily in the hands of two men—Secretary of State for Air William Weir and Minister of Munitions Winston Churchill—who were determined to rationalize aircraft production and field "clouds of aeroplanes." They also dramatically expanded the labor force at aviation factories by employing women and boys, and by November 1917, the British aviation industry was the world's largest. The aircraft firms increased their deliveries from 6,633 in 1916 to 14,382 in 1917. However, British engine deliveries lagged more than 3,000 units behind airframes in 1917, and domestic copies of the Hispano-Suiza were markedly inferior to the original, forcing the importation of nearly 5,000 engines, primarily from France.

Italy, the least of the Allied powers, delivered 3,861 airplanes and 6,276 engines in 1917 through a well-organized system of production and development that focused on 300-horsepower Fiat six-cylinder inline engines to power a variety of reconnaissance-bombers and flying boats. Bringing up the Allied rear, the United States—the new associate power that had entered the war in April 1917—was undertaking a confused and chaotic mobilization to fulfill grandiose expectations. It would best be counted on to supply pilots and spruce for aircraft construction.

In 1917 the German approach of developing specialized types of aircraft and appropriately training for specific operations, such as ground attack, proved more effective and efficient than the British method of simply throwing untrained Camel pilots into ground attack. Certainly, the military control of production and the use of airplanes was more unified and less politically unstable than in Britain or France—and it needed to be, given the shortages of materials and manpower in Germany.

Everywhere, however, the evolution of airpower demonstrated the signal importance of aero engines. The engine was the heart of the airplane. The crisis in aviation in France and Britain was fundamentally one of engine production, as both relied on increased engine power rather than aircraft design to provide the essential margin of difference for aerial superiority. Britain's magnificent Rolls-Royces, however, proved too complex for mass production. France displayed the best combination of quality and quantity. Germany, mired in material shortages, could neither move beyond the six-cylinder in-line nor increase its production sufficiently, no matter how reliable and durable its engines proved to be. The air war itself had become a full-scale war of attrition; consequently, the mass production and training of aircrews would prove decisive in 1918, the final year of the great conflict.

At the beginning of 1918, as the German air arm prepared for the March offensive on the Western Front, the army and navy continued their airplane and zeppelin raids, respectively, on England. By the end of May, both strategic campaigns had ended: the navy's, with the death of the airship commander and the end of airship production; the army's, with its commitment of all remaining bombers to provide support at the front. The strategic air campaign had failed in its first and most grandiose aim—to drive Britain from the war. The giant plane raids had caused the diversion of significant British fighter forces to home defense, and the tons of bombs they had dropped on Britain had caused death and destruction but only a limited disruption of production.

On the battlefront, the inadequate replacement of personnel and materiel limited the effectiveness of the German air force. The limited supply of materiel could be offset only by a marked qualitative ascendancy. Fortunately for German aircrews, some of the planes they flew in 1918—such as the high-altitude Rumpler C7 reconnaissance plane equipped with a 260-horsepower, high-compression Maybach engine, and the Fokker D7 fighter equipped with a BMW 185-horsepower engine—were superlative, their performance unequaled by their Allied opposition. Yet the fundamental problem remained: the aircrew training schools and the aviation industry could no longer meet the needs of the front. Some German fighter units continued to inflict higher losses on the enemy than they incurred up to the end of the war. Nevertheless, suffering irreplaceable losses of personnel and materiel, and dogged by stringent fuel rationing and the lack of material even for aircraft repairs, they neared exhaustion as the war drew to a close.

Ultimately, Germany's manpower reserves fell far short of the Allies', as did production and supply, the other key factors in a war of attrition. Deliveries of airplanes and engines fluctuated wildly, but neither attained the goals stipulated in the mobilization programs, and neither exceeded 1,500 units a month. Industry, beset by shortages of metals, fabrics, precision machines, tools, and coal and disrupted by transportation crises and labor unrest, had no prospect of dramatically increasing production. Production proved to be the Achilles' heel of the German air effort. Shortages had halved production during the winter of 1916–1917; now a near-total collapse of aviation production loomed in the winter of 1918–1919 owing to coal, material, and food shortages, had the war continued. Austro-Hungarian aviation merits only a footnote, as its total production for the ten wartime months of 1918 was only 1,989 planes and 1,750 engines—barely enough to keep pace with attrition. Its aviators paid for this inadequacy with their lives.

The German air force, although it fought to the end, was overwhelmed in the air; it had lost the war of aerial attrition to the French and British, aided by their lesser aerial allies. The French aviation industry produced 24,652 airplanes and 44,563 engines in 1918, overwhelming both its opponents and allies. The French air arm emphasized the employment of aviation in 1918 and trained increasing numbers of aircrews to man its fighter and bomber squadrons. This concentration on tactical aviation, and on the continued improvement of a limited number of fighter and bomber types and their engines, enabled both relatively high production at home and effectiveness over the front. In air strength, the French possessed the world's largest air force in 1918, and had the war continued into 1919, its new fighter aircraft and engine types would have portended increased difficulty for the German air arm.

The British aircraft industry delivered some 32,000 airframes but only 22,000 engines in 1918, requiring the importation of some 9,000 engines, primarily from France. The Royal Air Force, which attempted strategic bombing beyond the tactical air war on the Western Front, was widely dispersed on all the imperial fronts. The Italian aviation industry manufactured 6,488 airplanes and 14,840 engines in 1918, powering the Italian military and naval air arms to success against their dwindling Austro-Hungarian opponents. The United States' air service and aircraft industry would have offered the Allies a powerful air weapon had the war continued into 1919 and 1920. But the German army high command could read the handwriting on the wall in October 1918; they had lost the war, on the ground and in the air.

Postwar

In Germany, preparations for postwar military procurement and civil aviation had begun in 1917, as military and civilian agencies planned to promote commercial aviation to compensate for the impending postwar reduction of military contracts and to assure Germany a leading position in postwar aviation. The collapse of the imperial German government on 9 November 1918 and the formation of a provisional government the next day created an uncertain and tense atmosphere. As the spontaneous sprouting of workers' and soldiers' councils around Germany haunted onlookers with the specter of imminent revolution, the air force and aircraft industry groped for survival.

The armistice of 11 November 1918 prescribed the immediate demobilization of the German air force and the surrender of 2,000 military airplanes, particularly Fokker D7s and all night bombers. By 12 December, the Allies had received only 730 usable airplanes, and the German army, which still possessed some 9,000 airplanes after the surrender, simply refused to cooperate with the victors. Although the civilian government established an Air Office in the Office of the Interior to help the army liquidate military aviation, neither the *Kogenluft* nor the War Ministry brooked any interference from civilian agencies in the liquidation of military aviation.

When the *Kogenluft* was dissolved on 21 January 1919, Lieth-Thomsen became head of the War Ministry's aviation department, where he attempted to utilize modified military aircraft to form airlines and a military air courier service, ventures that he intended to use to maintain military readiness. Such enterprises actually competed with the plans of the civilian government and aircraft industry to develop civil aviation, leading to an "internal air war" between military and civilian agencies.

Meanwhile, the German aviation industry foundered in an inescapable dilemma: the absence of demand in all industrial spheres meant that neither aircraft manufacture nor conversion provided profits. In this atmosphere, Anthony Fokker fled back to Holland, smuggling airframes, engines, and tons of parts and equipment on six trains. Other aircraft firms collapsed.

The exception to the rule was Hugo Junkers, creator of the groundbreaking all-metal airplane. His accumulation of capital, plant, and experience enabled him to inform his design and engineering staff that the company would now concentrate on civil air transportation. In the spring of 1919 Junkers delivered all-metal fighters to the army and developed the prototype F13, an all-metal six-seater cabin monoplane that would become the most widely used transport plane in the world in the 1920s.

The disclosure of the terms of the Treaty of Versailles on 8 May 1919 ended Germany's limbo disastrously. Five articles—numbers 198 through 202 in Part 3 of the treaty—devastated German aviation, forbidding and demobilizing military and naval aviation, ordering the surrender of materiel, and prohibiting the production and importation of aircraft and parts for six months after the treaty took effect. The signing of the treaty on 23 June 1919 sounded the death knell of the German air force and aircraft industry of World War I. By the time the treaty became effective on 10 January 1920, the old guard of Germany's first air force had retired. General Hans von Seeckt, chief of the German Troop Office (the old General Staff in disguise), proclaimed the official dissolution of the German air force on 8 May 1920 with the epitaph, "The arm is not dead; its spirit lives."

Seeckt was well aware of how alive military aviation was, and not merely its spirit. The central occurrence of 1920, however, was the dismantling of German military aviation. The Inter-Allied Control Commission, created by Article 210 of the treaty to ensure adherence to its provisions, monitored German efforts to evade the treaty's restrictions, including the continued production of aircraft and the efforts of the Ministry of Transport's Aviation Department, under Captain Ernst Brandenburg (a famous bomber pilot during the war), to use German civil aviation as a vehicle to preserve military aviation.

The Allies continued the prohibition on German aircraft manufacture into 1921, by which time most of the industry, including many prominent firms, had collapsed or ceased production. The Inter-Allied Control Commission confirmed that practically all the old firms, if they had survived at all, were producing other implements. By 1921, the World War I German aircraft industry was no more.

The continued aviation restrictions, however, ultimately benefited the German army. The German Defense Ministry's secret rearmament in the Soviet Union enabled it to evade the total ban on military aviation, and the limitations on commercial aviation forced the few remaining German aircraft firms, such as Junkers, Albatros, and Heinkel, back into the arms of the army to secure contracts. When the army approached Albatros, Heinkel, and the new Arado company in 1922–1923 for the Russian venture, Junkers was already building an aircraft and engine plant in the Soviet Union for the army.

Although the deficiencies of these companies forced the army to rely temporarily on Anthony Fokker in Holland for airplanes, the connection between the army and the aircraft industry was reestablished for the future. The Allied bans had restored the symbiotic relationship between military and industry. The restrictions had severely crippled the German industry and the air force, but the few firms that survived and the new ones that arose in the 1920s were invariably and indissolubly linked to the German army in the clandestine rearmament of the Reich under the Weimar Republic.

The dissolution of the Austro-Hungarian monarchy ensured that the postwar reduction of its aviation industry would be disastrous. In Austria, the government asked the factories to continue production without profit to assure their workers of wages for six months, but attempts to sponsor civil aviation failed, and with no prospective contracts, the firms either collapsed or turned to other pursuits. The few tiny aircraft factories in Hungary did not survive their nationalization by Bela Kun's evanescent communist regime.

By 1921, the postwar contraction and demobilization, combined with the collapse of governments in central and eastern Europe, essentially terminated the existence of the air forces and aviation industries of the First World War. Individual firms and designers and young military and naval aviators would survive, some of them long enough to fight—and lose—another war twenty years later.

Conclusion

Overall, and at the risk of oversimplification, certain powers passed the test of aviation mobilization better than others in the Great War of 1914–1918. The French were most successful because they recognized the importance of the airplane, and particularly the engine. France mobilized aviation first and most extensively, using standardization on a few airframe and engine models and relying on manufacturers' increase in engine power to improve aircraft performance. The Germans made highly effective use of limited resources, but the excessive expectations for the zeppelin and the continued construction of airships, particularly after their disasters on the Western Front, represented a wasteful expenditure of resources. The War Ministry's inspectorate also delayed too long in drawing other automobile companies into aero-engine production to counter Daimler's near monopoly. German reliance on the six-cylinder engine, though limiting, was at least understandable in light of shortages of material and skilled labor.

The Central powers certainly lost the manufacturing race. The cliché that Germany was tied to the corpse of Austria-Hungary has a ring of truth in terms of aviation, as the dual monarchy's industrial weakness ensured its dependence on Germany. Germany also had to supply the Ottoman Empire with aviators, ground crews, and aircraft to maintain a small Turkish air service during the war. In contrast, among the Allied powers, France, for example, could secure the Hispano-Suiza engine in neutral Spain and then use its own automobile industry to produce it in numbers sufficient to shore up British production and even equip an American air service. Allied aviation superiority became crushing as the war went on. The loss of the manufacturing race on the ground, in the factories on the home front, meant that, all other things being equal, the longer the war continued, the more likely it was that Germany and Austria- Hungary would lose the war in the air, and they did.

Suggestions for Further Research

Works in English on Austro-Hungarian aviation and on aviation engines, despite the latter's tremendous importance, remain few or non-existent. The release of documents from former East German and even Soviet archives should provide additional insights into the development of the German air service and aviation industry.

Detailed studies of the air policies of the high command with regard to operations, manufacturing, and the allocation of economic priorities could be useful, as well as research into the training of aircrews and mechanics and the location and construction of airfields and hangars.

The tendency of English-speaking authors to concentrate on British and German aviation during the First World War has resulted in much attention to German airpower. Authors such as Von Hardesty and Robin Higham are ensuring that Russian aviation is paid more heed, but the absolutely crucial development of the French air arm and aviation industry and that of its Italian and Austro-Hungarian counterparts merits much more study.

Recommended Reading

My works on German and Austro-Hungarian military aviation—*Building German Air Power* (Knoxville: University of Tennessee Press, 1976) and *German Air Power in World War I* (Lincoln: University of Nebraska Press, 1982)—and on World War I aviation—*The Great War in the Air: Military Aviation from 1909 to 1921* (Washington, D.C.: Smithsonian Institution Press, 1993)—are essential reading on the subject. All three works contain extensive footnotes and bibliographies that can lead interested readers to other sources. An older study of early German airpower, John R. Cuneo's two-volume work *Winged Mars* (Harrisburg, Pa.: Military Service Publishing Co., 1942, 1947), continues to have merit. Also see Charles H. Gibbs-Smith, *The Aeroplane: An Historical Survey of Its Origins and Development* (London: HMSO, 1960), 66.

Also see Jürgen Eichler, "Die Militärluftschiffahrt in Deutschland 1911–1914 und ihre Rolle I den Kriegsplänen des deutschen Imperialismus," *Zeitschrift für Militärgeschichte* 24, no. 4 (1985) and no. 5 (1986).

Douglas H. Robinson's studies of zeppelins—*The Zeppelin in Combat: A History of the German Naval Airship Division, 1912–1918* (London: Foulis, 1962) and *Giants in the Sky: A History of the Rigid Airship* (Seattle: University of Washington Press, 1973)—examine in detail the exploits of the giant airships and their intrepid crews.

On the all-important topic of aviation engines, see J. A. Gilles's essential study, *Flugmotoren 1910 bis 1918* [Aviation Engines, 1910–1918] (Frankfurt am Main: Mittler und Sohn, 1971), and some excellent overviews: Herschel Smith, *A History of Aircraft Piston Engines* (Manhattan, Kans.: Sunflower University Press, 1986 [1981]), and Kyrill von Gersdorff and Kurt Grassmann, *Flugmotoren und Strahltriebwerke* [Aviation Engines and Turbojets] (Munich: Bernard and Graefe, 1981). Peter Gray and Owen Thetford's book *German Aircraft of the First World War* (London: Putnam, 1962) remains an indispensable reference work on German planes.

A number of works written in the early 1920s continue to have merit for their often semiofficial accounts of the air war from the German perspective. Among them are Georg Paul Neumann's collection, *Die Deutschen Luftstreitkräfte im Weltkriege* [The German Air Force in the World War] (Berlin: Mittler und Sohn, 1920); Hans Ritter, *Der Luftkrieg* [The Air War] (Leipzig: Verlag von Köhler, 1920); and Hans Arndt, "Der Luftkrieg" in *Der Weltkampf um Ehre und Recht* [The World Struggle for Honor and Justice], ed. Max Schwarte (Leipzig: Alleinvertrieb duch Ernst Finking, 1922), 4:529–651. More recent are Kriegswissenschaftliche Abteilung der Luftwaffe (KAdL), *Die*

Militärluftfahrt bis zum Beginn des Weltkrieges 1914, 3 vols., 2nd rev. ed., ed. Militärgeschichtliches Forschungsamt (Frankfurt am Main: Mittler und Sohn, 1965–1966), and Richard Blunck, *Hugo Junkers. Ein Leben für Technik und Luftfahrt* (Düsseldorf: Econ-Verlag, 1951).

Writings on Austro-Hungarian military aviation are few, but two indispensable studies are Ernst Peter, *Die k.u.k. Luftschiffer- und Fliegertruppe Österreich-Hungarns, 1794–1919* [The Austro-Hungarian Airship and Flying Troops, 1794–1919] (Stuttgart: Motorbuch Verlag, 1981), and Peter Schupita, *Die k.u.k. Seeflieger: Chronik und Dokumentation der österreichisch-ungarischen Marineluftwaffe 1911–1918* [The Austro-Hungarian Naval Aviators: Chronicle and Documents of the Austro-Hungarian Naval Air Service 1911–1918] (Koblenz: Bernard und Graefe, 1983).

On the French background, see Patrick Facon, "L'armée française et l'aviation 1891–1914" (paper presented at the Southern Historical Association meeting, 1985); Charles Christienne et al., *Histoire de l'aviation française militaire* (Paris: Charles Lavauzelle, 1980); Albert Etévé, *Avant les cocardes; les Débuts de l'aéronautique militaire* (Paris: Charles Lavauzelle, 1961); Antoine Odier, *Souvenirs d'une vielle tige* (Paris: Fayard, 1955). On the engine and airframe industries, see James Laux, "Gnome et Rhône: Une firme de moteurs d'avion durant la Grande Guerre," in *1914–1918, L'Autre front,* Cahiers du "Mouvement Social," no. 2 (Paris: Les Éditions Ouvrières, 1977); and Emmanuel Chadeau, *L'Industrie aéronautique en France 1900–1950: de Blériot à Dassault* (Paris: Fayard, 1987).

On the history of the Royal Air Force, the following are useful: Randolph S. Churchill and Martin Gilbert, *Winston S. Churchill,* vols. 2–4 (Boston: Houghton Mifflin, 1967, 1971, 1975); Harald Penrose, *British Aviation: The Pioneer Years 1903–1914* (London: Putnam, 1967); Malcolm Cooper, *The Birth of Independent Air Power: British Air Policy in the First World War* (London: Allen and Unwin, 1986); Walter Raleigh and H. A. Jones, *The War in the Air,* vols. 1–6 (Oxford: Clarendon Press, 1922–1937); and W. J. Reader, *Architect of Air Power: The Life of the First Viscount Weir of Eastwood 1877–1959* (London: Collins, 1968).

See also B. P. Flanagan's translation of "History of the Ottoman Air Force: Reports of Maj. Erich Serno," *Cross and Cockade* 2, no. 2 (1970): 97–145; no. 3 (1970): 224–43; and no. 4 (1970): 346–69, and Hilmer von Bülow's *Geschichte der Luftwaffe* (Frankfurt am Main: Moritz Diensterweg, 1934).

CHAPTER FIVE

Downfall of the Regia Aeronautica, 1933–1943

Brian R. Sullivan

It is only a slight exaggeration to state that the defeat of the Regia Aeronautica (Royal [Italian] Air Force) actually took place when Italy entered World War II. When Mussolini announced that he was leading his country into conflict alongside Germany against Britain and France on 10 June 1940, he unwittingly placed his air force in a hopeless situation. Clearly, the French were already beaten. But Il Duce's belief that the British would also collapse before the might of the Germans over the next weeks or months would prove as mistaken as his recently abandoned conviction that war between the Axis and the democracies would not begin for years.

Based on Mussolini's strategic directives from late 1938 until early 1940, the Regia Aeronautica leadership anticipated a peaceful immediate future and prepared their service accordingly. Even so, the regime, the air industry, and the air force staff had already made a series of disastrous misjudgments about economics, international trade, manufacturing policy, labor management, air-war doctrine, scientific research, technology development, armament acquisition, training, and interservice cooperation. These decisions created near-insurmountable barriers to a Regia Aeronautica victory in a major war against a powerful industrial-technological state. It seems highly unlikely that so many major mistakes could have been corrected in the course of a few years.

Thus, whether Fascist Italy entered a world war in 1940, 1943, or even 1945, a ruinous outcome would have resulted, whatever the differences in the details. Given the weaknesses under which the Italians would have fought any war, even a German victory would have been an Italian defeat.

Nonetheless, in the actual war they fought, Italian airmen exhibited courage, tenacity, ingenuity, and a measure of nobility. Although it would be false to depict the Regia Aeronautica as brilliantly successful in 1940–1943, it did not wage war nearly as badly as is commonly supposed. The problems alluded to earlier, along with constraints imposed by geographic and social realities, placed enormous burdens on the combat effectiveness of the Italian air force. Indeed, the salient reality that emerges from a study of the Italian air force in the Second World War is not that it was finally demolished by British and American airpower. Instead, that the Italian airmen did as well as they did for as long as they did under extremely trying circumstances—sometimes demonstrating admirable effectiveness, and always displaying outstanding heroism—stands out as their fundamental achievement between June 1940 and September 1943 and deserves a degree of grudging admiration. An examination of the impediments under which they struggled makes clear, however, that they were almost certainly doomed from the start.

Decades before the outbreak of World War II, the Italian air force already suffered from significant weaknesses that were not easily correctable. Perhaps most serious was the incongruity of doctrine and practice, a problem that would plague the service until mid-1941. Mussolini's personal enthusiasm for flying; the early adherence of Giulio Douhet to the Fascist movement two years before his *Il dominio dell'aria* (Command of the Air) appeared in 1921; and the Fascist cults of daring, youth, romantic death, and modernity—all embodied in aerial warfare—together virtually ensured the foundation of the Regia Aeronautica as an independent service in March 1923. Douhet declined Mussolini's offer to head the new force, recognizing that his fiery nature and propensity toward polemics would make his leadership problematical. But Douhet's theories were accepted unreservedly by Mussolini; Il Duce never succeeded in imposing them on the army and navy, but they permeated the Italian air force from its inception and thrust its early development in highly impractical directions.

Severe interservice rivalry, particularly with the politically powerful army, combined with a paucity of funding for all the armed forces and the weaknesses of the Italian aircraft industry, made even an attempt to realize Douhet's concepts impossible. Throughout its first decade, the Regia Aeronautica received an average of 13.4 percent of the total armed forces budget—half that of the navy, and a quarter that of the army. Such allocations were hardly consonant with Douhet's eventual insistence that its sister services be reduced to border and coast guards and that the air force be expanded into a huge bomber force, capable of obliterating enemy cities in a few hours with incendiary and gas bombs. But the air force did stress the acquisition of bombers over other types of aircraft, acquired an array of chemical weapons and tested them on rebellious Bedouin in Libya, willingly ceded ground air defense of civilian targets to the incompetent Fascist militia, and gained Mussolini's agreement to provide only minimal air support to the army and navy.

The 1930s

From 1926 to 1933, under the charismatic leadership of the powerful Fascist leader Italo Balbo, the Regia Aeronautica gained national adulation and international acclaim. Its pilots and Italian aircraft designers set numerous records, while Balbo himself led a series of trans-Mediterranean and trans-Atlantic squadron flights designed for both publicity and intimidation. But Mussolini never gave Balbo the funds he demanded for the development of long-range strategic bombers and a fourfold expansion to achieve the dreams of his friend Douhet. Threatened by his air minister's popularity following Balbo's squadron flight to Chicago in 1933, Il Duce exiled him to Libya as governor later that year and took over as minister himself. Mussolini appointed General Giuseppe Valle as air undersecretary to actually run the Regia

Aeronautica, then made him chief of staff as well in March 1934. Valle held both posts until November 1939, but he lacked Balbo's independent power base. Beholden to his master, anxious to fulfill Il Duce's wishes, fiercely opposed to interservice cooperation, and determined to wage an independent strategic air war, Valle provided Mussolini with an increasingly dishonest portrait of air force realities. Meanwhile, the incongruity of Regia Aeronautica's doctrine and strategic planning with its operational capabilities worsened.

As Europe grew increasingly unstable after the Nazi seizure of power in January 1933, Mussolini greatly expanded military spending, both to defend against the German menace and to take advantage of opportunities for Italian expansion. During Valle's tenure, air force funding soared both arithmetically and in proportion to the other services. From mid-1936 onward, the air force received more money than the navy, with the gap widening thereafter. The seven annual air budgets of 1933–1940 averaged 24.5 percent of all military spending. (In comparison, the Luftwaffe received 36 percent of German military spending in 1934; the Royal Air Force received 38 percent of the British military budget for 1938 and 41 percent the following year.) But as an indication of growing Regia Aeronautica strength, the 1933–1940 figures are deceptive. Roughly two-thirds of those funds were abortively applied to bolster the Austrian and Hungarian air forces; to support the conquest of Ethiopia and the subsequent protracted counterinsurgency in East Africa; to back the Italian intervention in the Spanish Civil War and help the Nationalist air force; and to facilitate the invasion of Albania. Mussolini's acquiescence in the *Anschluss* in March 1938, Hungary's submission to German dominance, the strategic irrelevance of Spain and Italian East Africa in World War II, and Mussolini's May 1939 alliance with Hitler meant that most of the 1933–1940 air force spending had been squandered.

Such fruitless resource allocation and intense Regia Aeronautica involvement in the Italian–Ethiopian and Spanish civil wars proved highly injurious to the air force. Funds that might have purchased large numbers of technologically advanced aircraft and equipment were used to pay for these operations instead. Combat success against inferior opponents in East Africa and Spain led to overly optimistic conclusions. Bellicose Fascist propaganda made objective analysis difficult. Legislation that had created totally separate testing and purchasing offices within the air staff encouraged mistakes based on these activities.

In August 1936, three months after the official end of the Ethiopian war and shortly before major Italian participation in the Spanish Civil War began, the air force General Staff undertook studies for a major expansion and modernization of its air fleet. Program R was to be completed by the spring of 1938. The goal was a force of some 3,170 aircraft, almost all to be ready in the second half of 1940, except for 18 four-engine heavy bombers that would not come on line until February 1941. Some 278 would be stationed in Italian East Africa and 426 assigned to the navy. Of the remaining 2,448 aircraft, 966 would be medium bombers, 911 fighters, 338 reconnaissance planes, 132 ground attack and dive-bombers, 65 colonial, and 36 long-range transports. Many of these aircraft types were already in series production or available as prototypes. But some 750 of the fighters were to be of totally new design and constituted the most ambitious segment of Program R.

Aircraft Design and Selection

Shortly after this decision, the Air Ministry announced a competition for a modern interceptor, and the air force held trials for six entrants in early 1938. Based on results, Valle instructed the Office of Engineering and Aeronautical Construction to acquire the Fiat G.50 and the Macchi-Castoldi MC.200. Both all-steel warplanes enjoyed high maneuverability, sturdiness, and good handling; the MC.200 also provided extraordinary visibility from its hump-mounted cockpit, as well as superb climbing and diving ability. But the weak 840-horsepower Fiat A74 engine powering both allowed a top speed of only 293 and 314 miles per hour, respectively, and an armament of just two 12.7 mm machine guns. Furthermore, to save weight, the two fighters carried radio receivers but no transmitters. Still, many pilots and their patrons on the air staff preferred slower but even more maneuverable lightweight biplanes capable of agile aerobatics in dogfights. Ongoing successes in Spain with such an aircraft, the Fiat CR.32 (the Italians eventually downed about 500 Republican warplanes, including many Soviet Polikarpov I-16s, while incurring aerial losses of only 86), provided strong arguments for the proponents of such fighters. However, others on the air staff and a good many pilots realized that the fabric-covered biplane had reached the apogee of its potential. Meanwhile, all-metal monoplanes with powerful engines and armed with multiple machine guns or cannon had already surpassed them in performance. More advanced models promised far greater capabilities. The Germans had introduced the Bf 109B into combat over Spain in April 1937, and it proved remarkably successful over the next six months. Valle's staff soon became cognizant of the German fighter and its technical advantages, but their hearts and those of most fighter pilots lay elsewhere.

By mid-1938, Italian and German fighter pilots in Spain, flying CR.32s and Bf 109Bs and Cs, respectively, had fought together in a number of operations. The relative merits of these two types of aircraft had already provoked a lively debate in the pages of the official air force journal *Rivista Aeronautica,* reflecting similar arguments within the air staff. As a result, Valle issued requirements for a more advanced fighter than the G.50 and MC.200, but trials of six new interceptors were all disappointing. The Testing Office report stressed the superiority of the Bf 109 to all the Italian prototypes and urged the development of better ones from which to select a single Regia Aeronautica fighter for the mid-1940s. Valle agreed. However, the manufacture of both the G.50 and the MC.200 series had proceeded very slowly, especially that of the latter. As a stopgap, the chief of staff insisted that the Aeronautical Construction Office order 200 of the Fiat CR.42, a mixed steel and fabric-covered biplane with fixed undercarriage that had not been entered in the second fighter competition. The Construction Office falsely described the CR.42 as only an improved model of the CR.32. The major loser in the fighter trials, the Caproni group, was appeased by a few orders for its new aluminum monoplane, the Re.2000. Based on the Seversky P-35 (its American manufacturer needed cash and apparently sold the plans without obtaining an export license), the Re.2000 had been developed for the export market. When tested by the Swedes and Hungarians who purchased it, the Re.2000 proved superior to the Bf 109E in close combat, although inferior in armament and thirty miles per hour slower.

Until mid-1939, the air force continued to buy the CR.32 to supply both Italian fighter units and the Nationalists in Spain. The limited production capacity of the Italian aviation industry was divided among four fighters, one of which was already obsolete, another obsolescent, and two of rapidly fading modernity. The ugly duckling of the lot, the Re.2000, proved superior in combat to the G.50 and especially to the CR.42. Thanks to its elliptical wing, the Re.2000 did not spin when subjected to the aerobatics favored by Italian pilots, a tendency that proved fatal to many who flew the G.50 and MC.200 for the first time. Moreover, unlike the spruce and special steel required to build those two warplanes—materials that were in short supply in Italy even before World War II—the aluminum required to construct the Re.2000, as well as hydroelectric power to extract it from bauxite ore, was relatively abundant in Italy. Finally, thanks to its American-inspired design, the Re.2000 easily lent itself to mass production.

The CR.42 proved equal to the Gloster Gladiator, which the Royal Air Force (RAF) continued to employ in the Mediterranean and Africa in 1940–1941. Flown by a skilled and aggressive pilot, the Fiat biplane with its superior maneuverability could even offset the speed and firepower of the Hawker Hurricane. In 1939 it impressed the Belgians, Swedes, and Hungarians sufficiently that they ordered a combined total of 180. The CR.42 would remain in production until 1943, by which time the RAF had long since replaced the Gladiator and most of its Hurricanes with Supermarine Spitfires. The G.50 and MC.200 were obsolescent when they went into series production in the fall of 1938. They would prove to be only a rough match for the Hurricane (designed in 1934) and distinctly inferior to the Spitfire, each already in service, with characteristics well known to the Italians, by the summer of 1939. Nor were the two Italian fighters equal to their French contemporaries, the Dewoitine D-520 and Arsenal VG-33. Officially, the Re.2000 was rejected because gas tanks in its wings made it vulnerable, but that could have been corrected. Valle's choices and his surreptitious acquisition of the Fiat CR.42 make it difficult to avoid the conclusion that corruption or politics tainted the selection process.

A somewhat similar story surrounds the three-sided competition in 1939 among the projected Cant Z.1014 and Caproni Ca 204 and the existing Piaggio P.108 for a contract to provide the air force with a heavy bomber. Based on the paper specifications presented by the manufacturers, the Z.1014 was judged best, followed by the Ca 204. But Piaggio had already built two prototypes of four-engine heavy bombers in 1937 and 1938 and had employed that experience to construct a prototype P.108 as a private venture in 1939. After Piaggio agreed to lower its price by half, it won the order. The P.108 did not enter service until 1941, and only forty-five were constructed, as all suffered from serious structural problems and unreliable P.XII RC35 engines. Many P.108s inexplicably failed to return from long-range missions, and the crash of another one killed Mussolini's favorite son, Bruno, in August 1941.

Did the fact that the Z.1014 and Ca 204 existed only on paper, while the P.108 was available for flight-testing and far closer to series production, influence the selection, or did bribery play a part in the process? Whether from choice or necessity, Valle seems to have tolerated such behavior.

Material Readiness, 1939–1940

Whatever considerations influenced the selection of Italian warplanes, when the Regia Aeronautica entered World War II in June 1940, nine of its twenty-one fighter groups and two of its six independent fighter squadrons were partially or fully equipped with the antique CR.32.

Most of the other units were flying CR.42s, supplemented by only a small number of MC.200 and G.50 fighters. In the spring of 1939, when Mussolini agreed to an alliance with Hitler, Il Duce had insisted, and the Führer had agreed, that the European Axis would not begin a war against the democracies until 1943 at the earliest. Italian strategic planning and military preparations largely proceeded according to that schedule. For the air force, this expectation encouraged continued low production levels of many outmoded aircraft and engines and a leisurely approach to designing new prototypes. But Hitler provoked the Second World War in September 1939. Though enraged by this disregard of Italian interests, Mussolini remained eager to enter the conflict, from which he expected considerable territorial booty. However, Il Duce reluctantly acknowledged his lieutenants' warnings that the army in particular, but also the air force, was woefully unprepared. Over the next few weeks, the dictator learned that the chiefs of staff had lied to him for years about their services' readiness for war. Mussolini forced both the army chief of staff and Valle into retirement on 31 October 1939. Il Duce appointed General Francesco Pricolo, commander of the Second Air Zone, which encompassed northeastern Italy fronting Yugoslavia, as the new air undersecretary and chief of staff.

Pricolo inherited an air force of some 5,200 aircraft, including 2,700 mostly outmoded warplanes. About 900 were so worn out that he ordered their immediate scrapping. Subtracting the 200 planes with training units, and given the chronic lack of spare parts and insufficient numbers of skilled ground personnel, the new air chief calculated that he had only 647 bombers and 191 fighters that were operational. The realization stunned and depressed him. During his first meeting on 18 November with the other chiefs of staff and Il Duce's chief military adviser, Marshal Pietro Badoglio, Pricolo gradually realized that the army was even worse prepared, and the navy was little better. Lack of fuel and lubricants stood out as the major concern for all, yet no central authority existed to control petroleum supplies and other essential materials. Each chief of staff was reluctant to reveal the true condition of his service, lest that lead to demands for resource sharing. Ironically, Pricolo was equally guilty of such reticence. What seemed clear was that until the Regio Esercito (Royal [Italian] Army) had completed its rearmament, it would remain on the defensive—except possibly in the Balkans, where opportunities for successful operations against the Yugoslavs might arise. Clearly, that contingency had influenced the choice of Pricolo to command the air force. Should war between the Axis and the democracies take place, the Italians would undertake offensive but uncoordinated air and sea operations. Presumably, as Germany's ally, Italy would receive territory from the defeated Western powers at the peace table. Pricolo set to work to ready the Regia Aeronautica for such eventualities as best he could.

In theory, such dispositions would allow the air force to pursue a Douhetian strategy. In practice, however, the lack of sufficient numbers of medium bombers, as well as the lack of any heavy bombers at all until 1942, left the air force facing a strategic void. From early 1939 until mid-1943, the great majority of Italian bomber units employed three aircraft models. Two, the Fiat BR.20 and the Savoia Marchetti 79, first saw service in Spain in 1937. These Italian bombers surpassed the early-model German He 111 and Do 17 in speed, range, and bomb load; however, they did so due to lighter, partially wood and canvas structures and inferior defensive armament, to compensate for their weaker engines. Valle ordered the development of a better medium bomber. In 1939 the Regia Aeronautica began receiving the Cant Z.1007 at the same time the Luftwaffe acquired its first Ju 88s. By then, superior German aircraft engines gave the Ju 88 the capacity to carry a much heavier bomb load, although over a shorter range and at a slightly slower speed than the Cant bomber. Furthermore, the Z.1007 was built of wood and, like the SM 79, needed three engines to equal the two that powered the German medium bombers. The advantages of the Italian aircraft over their German counterparts were more than offset by their deficiencies. The wood and canvas construction of the Italian bombers caused them to warp when exposed to moisture, offered no protection against machine gun and antiaircraft fire, and made them easily flammable. The need to mount a third engine in the nose greatly reduced the forward vision and accuracy of the bombardier in the SM 79 and Z.1007. In addition, the three Italian bombers lacked the power to carry the communications and navigation equipment that enhanced the performance of the He 111, the Do 17, and especially the Ju 88. These three German bombers were followed by a series of significantly improved models that were more powerful and better equipped; in contrast, the Italians continued to build their medium bombers with only minor design changes until early 1943.

Doctrine and Strategic Options

Geography imposed other obstacles to carrying out the strategy advocated by Douhet. Whereas Paris was only some 200 miles from the Rhineland across low-lying terrain, the French capital was 375 miles from the nearest Italian air bases in Piedmont, and London was 600 miles away. To reach either would have required a fuel-consuming flight over the Alps or a long detour around them. Given the limited range of the medium bombers available to the Regia Aeronautica in 1939–1940, delivering the devastating knockout blow against those enemy capitals proposed by Douhet would have been impossible. Finally, French airfields in Corsica were separated from Rome by only 160 miles, so Valle had ruled out any strikes against Paris for fear of devastating retaliation against the Italian capital, a worry that had weighed on him for years. Despite air force boasts that it dominated Italy's mare nostrum from above, these limits prevented the use of Douhet's ideas, even against the Balkan states.

Even sustained Italian air operations against military, communications, and industrial targets in the greater Mediterranean–Middle East region were limited by the range of Italian medium bombers. The French Mediterranean ports Bizerte and Malta could be attacked fairly easily by aircraft from Tripolitiania, Pantelleria, Sicily, and Sardinia, but Alexandria lay 350 miles from Italian airfields in Cyrenaica and Rhodes, severely limiting the bomb loads with which Regia Aeronautica warplanes could attack that Royal Navy anchorage. A distance of 550 miles separated the Suez Canal from those same Italian air bases. Gibraltar was 800 miles west of the nearest Italian air base on Sardinia, the Mosul-Kirkuk oil fields lay 850 miles east of Rhodes, and those in Kuwait and Bahrain were 1,000 to 1,200 miles distant from the airfields of Eritrea.

One alternative was an offensive aeronaval strategy based on close cooperation between the Regia Aeronautica and the Regia Marina, but Valle and the air staff rejected this option as antithetical to waging an independent air war. Neither the navy nor the air force high commands had ever been enthusiastic about acquiring aircraft carriers. Thus, a rough consensus was reached on that potentially contentious issue, at the price of crippling the striking power of the navy in World War II. But Valle also opposed any initiative that might lead to subordination of air units to navy operational command. In 1933 the navy and air force agreed to develop torpedo bomber squadrons. When the possibility of war with Britain arose in 1935, however, such units had not been formed. Air force attempts at alternative approaches failed. The Whitehead factory of Fiume had developed an excellent 450 mm aerial torpedo, yet the two services could not agree on how to divide the costs of producing it. In fact, the basic air staff objection arose from the fear that torpedo bomber squadrons would be incorporated into the navy. Even if the Regia Aeronautica retained control of such units, they could suffer heavy losses when forced to fly low and slow to launch torpedoes. Such operations would benefit the navy, but the air force would pay a heavy price. Foiled, Valle simply refused to resolve the issue with the navy. Finally, in 1938 the Luftwaffe ordered 500 aerial torpedoes from Whitehead and agreed to pay for the factory to construct them. In return for accepting the diversion of aerial ordnance to Germany, Valle obtained agreement that the Regia Aeronautica would receive some of the weapons. As later events would demonstrate, aerial torpedoes dropped by skilled bomber pilots were worth every *centissimo* of their cost and that of the aircraft shot down while launching them.

The ground attack planes and dive-bombers produced by Breda and Savoia-Marchetti in the late 1930s proved too dangerous to fly and were useless for naval warfare in any case. (Only in 1941 did the Regia Aeronautica receive 100 heavily used Ju 87 Stukas from the Luftwaffe, as well as an effective dive-bomber based on the Re.2000 design later that year.) Instead, until the summer of 1940, the air staff remained wedded to high-level horizontal bombing attacks on enemy warships, despite repeated failures during peacetime maneuvers. In a vain attempt to better the chances of damaging naval vessels, the air staff decided to increase the number of weapons dropped in naval air strikes by using small bombs of 30 to 50 kilograms. But light warheads would not penetrate the armor of cruisers, aircraft carriers, and battleships, and poor accuracy made hits on unarmored but highly maneuverable destroyers unlikely. Similar difficulties arose engaging heavily fortified land targets using the prescribed 50-kilogram bombs. Valle ordered 100-kilogram bombs for both types of missions, but these also proved too weak, and funds for additional bomb purchases had been exhausted. Soon after taking command of the Regia Aeronautica, Pricolo squeezed more funding from Mussolini and ordered the production of 250- and 500-kilogram ordnance. These took time to produce and deliver, however, while huge numbers of the smaller bombs remained in air force dumps to be employed ineffectively in 1940–1941.

Another air theorist, Amedeo Mecozzi, argued that Douhet had greatly exaggerated the possible results of grand strategic bombing and that, in any case, Italy lacked the resources to put such ideas into practice. Mecozzi instead advocated tactical air strikes by ground attack and dive-bombers. Valle tolerated rather than supported Mecozzi's ideas, while privately conceding that they were valid, and he ordered aircraft to apply these concepts, which enjoyed considerable success in Ethiopia and Spain. At the same time, the long-serving chief of staff exacerbated Regia Aeronautica bombing deficiencies by enshrining Douhet's theories as necessary for an independent air force, yet he failed to acquire the aircraft such theories demanded. Valle also insisted that tactical air support requests receive air staff approval. That enhanced the Regia Aeronautica's independence but prohibited timely intervention in land and sea battles. But only repeated failure in 1940 made that clear.

The strategic orientation imposed on the armed forces by Mussolini remained in effect after September 1939. This made sense due to the wretched state of the Regio Esercito, particularly its antiquated artillery. Il Duce ordered all three services to prepare for intervention in the war as quickly as possible. After receiving detailed reports on the state of the armed forces, however, Mussolini reluctantly accepted the summer of 1941 as a reasonable target date. By then, the air force would be barely ready and the navy nearly so, but the army would remain years away from acquiring modern artillery. So the Regia Aeronautica would provide Italy's major means of offense—a concept fully in keeping with Il Duce's exalted notions of airpower—but it would have to compensate for the army's lack of firepower by extensive close air support, diverting its slender resources from waging an independent strategic air war.

To a degree, Pricolo favored cooperation with the other services. Shortly after becoming chief of staff, he ordered the accelerated formation of a torpedo bomber squadron to be operational in July 1940 and followed by others. Nonetheless, he considered ground interdiction, naval air strike, even air defense and air superiority missions all secondary to the Regia Aeronautica's main purpose: terror bombing of enemy cities. Thus, the new chief of staff remained fundamentally a Douhetian. He continued to focus on a role for which the air force was not equipped, while neglecting training for missions that were better suited to the aircraft on hand or those slowly emerging from the factories. Pricolo failed to anticipate that the war would require close cooperation with the other services and with air defense units, although the latter deficiency did not become fully apparent until late 1942.

Italy's geographic position granted the country a high degree of protection against air attack from some directions, especially the north. The defeat of France in June 1940 and the withdrawal of the RAF from the Continent left the Turin-Milan-Genoa "industrial triangle," where the majority of the Italian air industry was located, relatively safe from bombing. Despite their possession of bombers with greater load capacity and range, the British faced the same geographic barriers to launching raids on northern Italy from England as did the Regia Aeronautica in the reverse direction. During the brief British occupation of Cyrenaica in early 1941 and access to Greek airfields in November 1940 to April 1941, RAF assets were too slender and were relinquished too quickly to allow it to launch significant assaults on Italy. The daring and effective naval air strike against Taranto in November 1940 offered no alternative operational model, given the paucity of Royal Navy carriers and their vulnerability to countermeasures if such an attack had been repeated. But the arrival of ever-increasing numbers of American strategic bombers in the Mediterranean after mid-1942, the permanent Allied possession of Cyrenaica beginning in November, and the conquest of northern Tunisia in April–May 1943 gave the British and the Americans both the means and the locations to carry out sustained bombing of Italy. Even if the Allies had forgone the invasion of Sicily and the peninsula, Italian cities would likely have suffered destruction from the air in the second half of 1943. Italian air defenses were almost nonexistent, and the Germans, facing their own difficulties, were unlikely to have provided much help.

Air Intelligence

The Italian Air Intelligence Service (Servizio Informazioni Aeronautiche, or SIA) was established in 1935 and remained small, relying on naval intelligence for radio intercepts, until early 1941. Even after air intelligence created its own signals intelligence unit, its limited capabilities forced the Regia Aeronautica to focus it on the Mediterranean region, leaving the navy to monitor more distant traffic. Nonetheless, SIA functioned largely autonomously, although it used a carabinieri (national police) unit for counterintelligence and personnel surveillance. The latter activity involved compiling dossiers on airmen and especially officers that detailed not only their character and professional abilities but also their loyalty and that of their family and friends to the Fascist regime. Mussolini carefully studied such information on field-grade and general officers, making high-level promotions accordingly.

Until the war closed most Italian embassies, SIA posted nine air attachés abroad. By April 1942, however, Italy had air attachés only in Germany, Japan, Hungary, Spain, and Argentina. Before and during the conflict, Italian air attachés reported regularly on German air activities. To that end, the senior air force officers posted to the Berlin embassy from 1933 cultivated relations with Hermann Göring. Despite those friendships, the Luftwaffe did not share sensitive information. To compensate, SIA established a widespread espionage network throughout Germany in the 1930s. It also placed agents in France, the Balkans, and the Middle East.

Even before forming its signals intelligence unit, air intelligence began conducting electronic warfare. Its transmitters jammed navigation signals from Malta, mimicking them to lure British pilots to Sicily. These operations were enhanced after the X Fliegerkorps arrived on Sicily in January 1941 and erected a Freya radar station overlooking Siracusa. Meanwhile, exposure to the superior British and German electronics spread dismay through SIA and eventually the whole air force. The Italians' inferior ground radios, the general absence of two-way aircraft radios until mid-1941 and their poor quality thereafter, the lack of aerial radar sets, and poor navigation aids combined to depress morale.

The Italian navy had developed a number of effective radar systems by mid-1941, but interservice rivalry, incomprehension by the leadership, and the weak Italian electronics industry deprived the Regia Aeronautica of the benefits. By late 1941, the Luftwaffe had supplied some ground-based radar sets, concentrated in Libya and Sicily. Although the Regia Aeronautica finally managed to acquire an understanding of radar and obtained enough sets to put that knowledge to practical air defense use by 1943, the odds it faced by then proved overwhelming.

Air force carabinieri investigated anti-Fascist activities and "defeatism," as well as espionage. They routinely read air force personnel mail; closely monitored photographic activity near air bases and factories, as well as from commercial aircraft; and watched foreigners connected to air activities. SIA also directed its agents abroad to report any spying against the air force. In early 1941 its carabinieri discovered the radio signals of a large British air intelligence group and succeeded in capturing the entire ring over the next six months. The British were unable to reestablish such a network.

The Regia Aeronautica proved less able to secure its ciphers. By late 1940, British cryptographers had broken the Regia Aeronautica high-grade cipher and that used by its maritime search aircraft. The former success, combined with effective photo reconnaissance, gave the RAF a great advantage in operations over the Mediterranean area in 1940–1941. The latter assisted the Royal Navy to reroute Mediterranean convoys and anticipate Italian naval movements. However, Italian air units on Sicily used landlines, so British forces on Malta did not share such advantages. The Italians changed their high-grade air cipher in late 1941 and again in May 1943, and both times it took British cryptographers months to break it. However, the British ability to read German Enigma traffic revealed far more about Axis air activities in the Mediterranean. In general, SIA proved effective, but it could not compensate for Regia Aeronautica's inferiorities, especially too few aircraft.

Industrial and Labor Policies

To a large extent, preparing the air force for the struggles of 1940–1943 was beyond Pricolo's ability. He, like many other supporters of the theories of Douhet, Trenchard, or Mitchell, conceived of air war in terms of gigantic bomber attacks that would break opponents quickly, both psychologically and physically. Instead, World War II's air war became one of attrition, an agony lasting years longer than even the seemingly interminable trench warfare of the 1914–1918 Western Front. The air forces in the subsequent global conflict required a growing flow of aircraft, engines, and aerial ordnance, yet these lay beyond the capacity of the Italian economy and aircraft industry to supply.

In the last four months of 1939, Italian manufacturers produced an average of 196 aircraft and 404 aero engines per month. In the second half of 1940, after strenuous efforts by the air staff and the aircraft industry, average monthly production rose to 308 planes and 602 engines; in 1941 the monthly production averaged 291 and 605, respectively. Airframe output then declined in 1942 and fell precipitously in 1943. Engine production finally grew in late 1942–early 1943 to about 1,100 per month. For more than two years, however, engine manufacturers had raised strenuous objections to allowing automobile makers to apply mass production techniques to aero-engine building. Since Fiat and Alfa Romeo manufactured both engines and automobiles, this opposition can be explained only by the jealousy and rivalry among the other industrial groups involved. In comparison, the British government imposed rationalization on its air industry. Monthly British aircraft production peaked at an average of 2,197 in 1943 and 1944. From the Italian figures, it is clear that the Italian aircraft industry, as it was organized, supplied, and staffed during World War II, possessed very limited production capacity, especially for the more complex warplanes. After a maximum effort, airframe output roughly tripled, but engine manufacturing only doubled until the tardy intervention of the auto industry. By then, however, due to a concentration on single-engine aircraft by the Italians, there was an excess of Italian-built aero engines, which were sold to the Luftwaffe.

Pricolo's efforts and Mussolini's orders had greatly increased the number of modern operational aircraft available to the Regia Aeronautica when Italy entered the war in June 1940. Bombers had increased to 995 and fighters to an even more impressive 574, although 300 of these were CR.42s. Subtracting the modern bombers and fighters in East Africa left Pricolo with about 977 bombers and 542 fighters in the Mediterranean area. In addition, the air force possessed a total of about 700 reconnaissance aircraft, including those serving with the navy.

Still, of all the factors leading to the crisis that afflicted Italian airpower in early 1941 and led to the collapse of the Regia Aeronautica in mid-1943, the most important was inadequate numbers of aircraft. Other problems, such as a lack of skilled manpower for ground and aircrews and inadequate raw materials and fuel, presented grave obstacles to deploying a more powerful Italian air force. But given the exemption of all Italian male university students from obligatory military service (some 135,000 in January 1942, when the Regia Aeronautica numbered only about 130,000, including 8,000 officers), Axis access to European resources between mid-1940 and mid-1943, and German production of synthetic fuels, such difficulties could have been completely or largely overcome. Until the Italian air industry was restructured to use such assistance effectively, however, the Germans would not provide those resources. And that did not occur until late 1942.

In the years immediately preceding Italy's entry into World War II, Italian airframe manufacturing coalesced into nine industrial groups, and air-engine production into four groups. Two of these, Fiat and Piaggio, built both airframes and engines. All constructed their products in relatively small batches using highly skilled labor; these methods were deeply rooted in Italian culture, encouraged by the Air Ministry, promoted by the regime's labor policies, and favored by manufacturers because the lack of sufficient machine tools had left them dependent on the small numbers of highly skilled artisans who dominated the air industry workforce. Semiskilled laborers could have been introduced into the system to expand production significantly only if they could have been put to work running specialized machine tools. Air industry employment had soared from 9,700 to 45,700 between 1934 and 1938, largely through the hiring of semiskilled workers. But this nearly fivefold increase in the labor force had resulted in only about a 50 percent increase in production. By spring 1943, this employment had expanded to 115,000, yet monthly aircraft production averaged only 266. Admittedly, by then, Allied bombing had severely disrupted Italian production. Nonetheless, without the necessary machine tools, the increase in the air industry's labor force had minimal results. Nor would running the small numbers of existing machines twenty-four hours a day, seven days a week have been feasible. With few or no replacement parts available, the machines would have burned out far short of the total hours of machine life that lower-tempo use would provide. Increasing the workweek beyond sixty hours, except under extraordinary circumstances, was rightly seen as counterproductive, a fact that Mussolini himself stressed.

Given such constraints, the ministry had promoted batch manufacture for several reasons: it allowed frequent improvements in design during the service of an aircraft model; it avoided the acquisition of large numbers of disappointing or obsolete designs; it allowed for operational experience with a variety of aircraft of the same type to determine the best design; it permitted orders for different models of the same type of aircraft to be distributed among several companies, ensuring their economic survival and continuous work for their artisans; and it created opportunities for ministry officials to solicit bribes and kickbacks. Since the Fascist regime had eliminated independent labor unions, outlawed strikes, and forced down wages and kept them low, manufacturers could employ highly skilled artisans quite cheaply. Thus, industrialists favored the batch system because it allowed them to avoid the expense of purchasing large numbers of specialized machine tools and instituting costly innovations in manufacturing techniques. As a result, the Regia Aeronautica failed to receive the aircraft it needed under the merciless demands of high-intensity warfare. For example, whereas construction of a Bf 109 required 4,500 man-hours, the markedly inferior MC.200 took 21,000 hours of artisan work to complete. After unexpected German military successes in April–May 1940 suddenly induced Il Duce to lead Italy into war, two and a half years would pass before pressure from the government and impending defeat would radically alter the Italian air industrial system. Even then, the standardized mass manufacture of airframes and engines by semiskilled labor in long production runs inside huge assembly plants—the essential methods by which large numbers of aircraft would be built in the 1940s—was still largely beyond Italian industrial capability.

Airframes and Power Plants

Making matters worse were the multiplicity of aircraft types and the failure to develop high-powered engines. After taking command of the air force in November 1939, Pricolo attempted to reverse Valle's predilection for buying several designs of each aircraft category. Nonetheless, the number of models employed by the Regia Aeronautica rose from nineteen in 1939 to twenty-nine in 1943. This resulted from a vicious circle also related to the lack of specialized machine tools, which would have allowed the industry to adopt mass-production methods for a few standard models. Faced with that deficiency and a growing demand for aircraft from training and operational units, the air industry chose not to force several groups to produce a particular type. Likewise, when newer models were accepted by the air force, ongoing needs persuaded both individual manufacturers and the Air Ministry to keep older production lines running, while creating new ones. Finally, artisans familiar with handcrafting an older aircraft could not adjust easily to a different design. The problem was exacerbated beginning in late 1942 when Allied bomber attacks began destroying aircraft factories, making the batch production of even outmoded models better than none at all. One result was that Fiat continued to turn out the CR.42 biplane until the September 1943 armistice, even after it had begun to manufacture the superb high-performance G.55 in early 1943. Only in February 1943 did Italian manufacturers finally agree to a drastic reduction in the number of models of airframes and engines to be built. And that came only after intense insistence from Pricolo's successor, General Rino Corso Fougier, himself the object of strong German pressure. Under the plan worked out between the air staffs, production would be concentrated on the smallest practical number of models chosen from the best designs of both countries. But by September 1943, implementation had just begun. Nonetheless, those preliminary steps allowed the Germans to profit from increased Italian air industry production, heavily concentrated in the Po Valley, in 1943–1945.

Contrary to appearances—including the speed records achieved by Italian aircraft, the mass long-distance flights led by Balbo, and the high quality of Italian automobile engines—the Regia Aeronautica entered World War II with aircraft powered by seriously deficient engines. In 1933–1934 Valle strongly encouraged the four aero-engine groups to abandon liquid-cooled, in-line engines for air-cooled, radial designs. At the time, the latter delivered more power for weight and seemed less likely to suffer catastrophic damage in aerial combat. Since the Italian air industry had little expertise with radial engines, it acquired licenses to build French, British, and American models. The foreign suppliers sold proven but dated engines. The experience the Italians had enjoyed with liquid-cooled engines was squandered, and the imposition of autarky further retarded Italian aero-engine development. Such policies restricted access to imported alloys and chemicals necessary to create engines capable of resisting higher temperatures and to make the high-octane fuel essential for enhanced performance. In turn, those restrictions discouraged attention to foreign technological advances based on such materials.

The Air Ministry inhibited other solutions to these problems by insisting on unified contracts for each phase of engine research, design, prototype construction, and testing; this discouraged innovation, a situation that was exacerbated by a lack of capital. Ministry insistence on 1,000-hour test runs for engines with a likely operational lift of 500 resulted in overly heavy engines.

Recognizing that Italian manufacturers would not produce an effective high-powered engine anytime soon, Valle obtained German agreement to license the Daimler Benz 1,175-horsepower DB.601A-1 to Alfa Romeo for Italian production and to sell fifteen examples directly to the Regia Aeronautica in January 1939. The DB.601 powered the Bf 109 and had been in series production since late 1937. But the aforementioned problems afflicting the air industry, including slow delivery of the necessary specialized German machine tools, delayed manufacture. Meanwhile, prior to his dismissal, Valle had turned over some of the German engines to the Macchi and Caproni groups. Pricolo accelerated production of the German engines, but by then, more than a year had been wasted. The DB.601s allowed the transformation of the MC.200 and Re.2000 designs into prototypes of the far superior MC.202 and Re.2001, which first flew in mid-1940. Macchi could bring its fighter into production far faster than Reggiane, so the former received the majority of the Italian-made DB.601s. The MC.202 finally entered service in November 1941. Engine deliveries never exceeded sixty per month, however; in contrast, 1941 German monthly production of the same engine averaged six times that. The same month the MC.202 entered combat, Fiat bought the license for the 1,465-horsepower DB.605, making possible the superb MC.205, G.55, and Re.2005 fighters. But because German industry itself was desperately short of machine tools at the time, Fiat could buy only thirty-seven directly from the Luftwaffe to begin production in November 1942. However, the German–Italian aircraft production agreement of February 1943 led to arrangements for Isotta Fraschini and Alfa Romeo to join in expanded Italian output of the DB.605. Of the three superb Italian fighters using such engines, however, only the MC.205 became operational in relatively significant numbers before the armistice.

The Command Structure

In the early 1930s, Mussolini had imposed a peculiar command structure on the armed services to enhance his personal power. Between July 1933 and October 1934, Mussolini assumed the portfolios of the three armed forces ministries, while appointing the three undersecretaries to be chiefs of staff of their services. In reality, each undersecretary carried out the duties of minister, while the deputy chief of staff performed the functions of his nominal superior. Since the six generals and admirals involved had to act in a capacity above their statutory powers, they needed Mussolini's approval for any significant action, giving Il Duce close control. On the one hand, he frequently imposed political decisions on his military leaders without recognizing that such directives were not militarily feasible. On the other hand, Mussolini's recognition of his lack of technical knowledge usually led him to defer to his senior generals and admirals on what appeared to be purely professional issues, even when he sensed that they were mistaken.

Despite Badoglio's title of *Capo di Stato Maggiore Generale,* there was no supreme General Staff due to both Mussolini's fear of such concentrated power and ferocious resistance to central control by the three services. The marshal simply presided over a small office staff to convert Mussolini's vague policy directives into general strategic guidelines for the three chiefs of staff, whom he could occasionally nudge toward consensus. Badoglio appreciated the expanding potential of airpower, thanks to the technological advances of the interwar years. Still, he remained profoundly attached to the tactical and operational concepts he had used to help guide Italy to victory over the Austro-Hungarians as deputy army chief of staff in 1918, over the Bedouin in Libya as governor-general from 1928 to 1933, and over the Ethiopians in 1936 as supreme commander in East Africa. In brief, he considered Douhet's theories nonsensical and considered the air force most useful in a ground attack or interdiction role. Yet for a man of supposedly limited intellect and imagination, the marshal possessed a surprising grasp of the value of aeronaval cooperation. By 1930, he had concluded that the navy should acquire aircraft carriers and did what he could to push the air force and navy toward formulating compatible doctrine and strategic plans. Badoglio's near-total failure in that regard, as the conflict over forming torpedo bomber squadrons illustrates, indicates the limits of his power.

This command system proved highly dysfunctional. The three military staffs made strategic plans in isolation from one another and from Badoglio. Each service chief adopted an obsequious attitude toward the dictator, desperate to please him and to hide anything that would not. Thanks to the particularly Fascist spirit that permeated the Regia Aeronautica, it attracted about 1,100 of Mussolini's relatives, lieutenants, and party officials to reserve commissions during Valle's tenure. A few of them proved to be capable leaders, but most were drones. Although they probably benefited the air force in terms of budgets and political influence, they also introduced a great deal of nefariousness and outside interference. Galeazzo Ciano, Il Duce's son-in-law, his foreign minister from 1936 to 1943, and a bomber pilot in the air force reserve, provided the most egregious example. Italian interservice rivalry was already ferocious, and such supposed favoritism subjected the air force to even more hatred and jealousy from the army and navy leadership.

Operations, 1940–1941

Strenuous efforts by Mussolini and Pricolo gave the Regia Aeronautica 2,300 modern operational aircraft when Italy entered World War II in June 1940. These included about 1,000 bombers, 700 reconnaissance aircraft, and 600 fighters, although some 300 of the last were CR.42s. For several months, the moderate tempo of air operations and the losses incurred proved sustainable. During the 11–24 June 1940 campaign against the French air force over the Alps, Corsica, Tunisia, and nearby waters, the Regia Aeronautica lost some 40 aircraft to all causes and destroyed about 70 French planes, mostly on the ground. For the next three months, the Italians concentrated on bombing Malta and Alexandria, flying maritime reconnaissance for the navy, and denying Libyan airspace to the RAF. On 9 July, about 75 Italian bombers attacked the British Mediterranean Fleet, but another 50 bombed their own warships from high altitude, following the naval battle off Punta Stilo, Calabria. These aircraft dropped 514 bombs. Save for one bomb that hit a British light cruiser, the Italians inflicted nothing worse than near misses on either side. In August, expecting a quick end to the war through a successful German invasion of England, Mussolini pressed a reluctant Hitler and an even more reluctant Pricolo to allow the 178-plane Corpo Aereo Italiano (CAI) to participate in the Battle of Britain. For a modest investment, Mussolini expected to share in the glory of Operation Sea Lion. However, inadequately trained in blind and instrument flying, CAI aircrew arrived in Belgium to discover flying conditions quite unlike those of the Mediterranean Sea and Libyan desert. As a result, they and their BR.20s, G.50s, and CR.42s accomplished little: 137 bomber sorties to drop 27 metric tons of ordnance, and 590 fighter sorties. These resulted in 9 losses inflicted and 20 sustained; accidents cost another 18 aircraft.

Meanwhile, Mussolini prodded a prevaricating Marshal Rodolfo Graziani—who had replaced Balbo as commander in Libya, following the latter's accidental death from Italian antiaircraft fire—to invade Egypt in early September. Il Duce also renewed plans to attack Greece, a conquest he had considered since early 1939 to dominate the Balkans. As Graziani's divisions approached Alexandria, Mussolini expected the Mediterranean Fleet to seek refuge in Greek ports. He would prevent that with a preemptive strike, as well as restore Italian influence in the Balkans, which had been damaged by recent German and Soviet advances in the region. After a large German military mission entered Romania in early October 1940 to train its army and to protect the Ploesti oil fields, Mussolini decided that he must act or suffer permanent German hegemony in southeast Europe. He hastily arranged an invasion of Greece that began on 28 October.

Pricolo concentrated about 380 aircraft for the campaign. Given the paucity of all-weather airfields in Albania, he based many of his bombers in Puglia, where the Regia Aeronautica enjoyed an extensive base system on the coastal plain stretching from Foggia to Lecce. The air chief did not expect significant resistance from the small Hellenic air force, which was equipped with 110 mostly outdated warplanes and another 130 reconnaissance and transport aircraft. Mussolini had ordered him to demolish every Greek town of more than 10,000 inhabitants, and Pricolo was determined to prove Douhet's theory that a country could be defeated through terror bombing. Italian bombers began carrying out such attacks on the first day of the invasion, but their mission soon underwent a radical alteration.

Heavy autumn rains began falling on the mountainous terrain through which the Italian forces advanced, slowing their invasion considerably. Soon thereafter, Italian army disorganization, poor leadership, and lack of artillery, combined with Mussolini's arrogant decision to attack numerically superior Greek forces, provoked disaster. The Hellenes halted the Italians, then began a counteroffensive. As the Italian army retreated, the Regia Aeronautica was ordered to the rescue. Despite being poorly trained to fly in bad weather, the Italian pilots were called on to bomb and strafe the advancing Greeks in lieu of the deficient Italian artillery. The fact that Albania had only three airfields with concrete runways meant that many air units had to take off and land in mud. Accidents and mechanical breakdowns multiplied. Aircraft were rushed to air bases in Puglia to support the campaign from better facilities. But flights across the foggy Adriatic proved dangerous for aircrews with little or no instrument flying experience and provided limited time over targets. Although Pricolo's bombers continued to raid Greek cities and towns, such attacks dwindled in frequency and intensity. The army's desperate need for reinforcements required the air staff to divert many bombers to transport missions. These dumped a chaotic assortment of unarmed infantry onto Albanian airfields, after which they were fed into the fighting haphazardly. As the Regia Aeronautica's losses mounted—more to damage in mishaps than losses in action from the small Greek air force and the smaller RAF contingent that joined it—Pricolo could not simultaneously sustain a massive troop airlift, terror bombing, interdiction strikes against Greek communications, and enough attacks on frontline enemy units to slow their advance. Mussolini begged Hitler for assistance, and soon thereafter, Luftwaffe Ju 52s arrived in southern Italy to assume much of the transport mission. By the end of the campaign, the Germans had flown 40,000 Italian troops across the Adriatic and evacuated 13,000 to Italy; the Regia Aeronautica had transported 31,000 men to Albania and flown out 8,000 wounded. Even so, by early December 1940, the Italian air force was barely able to sustain operations in the theater. Soon afterward, the Regia Aeronautica broke.

Graziani's divisions had dawdled a few miles inside Egypt for two months. Italian offensive operations from September to December were confined to those carried out by the 650 aircraft based in North Africa. But due to aggressive RAF patrolling and a range of deceptions, Regia Aeronautica reconnaissance flights did not detect British preparations for a counteroffensive. On 9 December a small but mechanized Commonwealth army struck back, supported by a well-trained but modest air force. Poorly placed Italian defenses and their lack of antitank guns to stop heavily armored British tanks allowed the attackers to break through and cut off Graziani's forward units. That defeat precipitated a series of Italian retreats, captured strongholds, and overrun airfields stretching across hundreds of miles of Cyrenaican desert. As Graziani's men either were cut off and forced to surrender or fled to avoid capture, most of their artillery fell into Commonwealth hands. As in Albania, the Regia Aeronautica was called on to substitute, subjecting the advancing enemy to what harassment it could. But the speed with which the Commonwealth advanced and the Italian ground forces fell back led to one Regia Aeronautica air base after another being abandoned.

Many damaged aircraft could not be flown out; precious fuel, ordnance, and spare parts were lost; and irreplaceable ground crews were taken prisoner. The rout of the Italian army swept up the air force as well. Pricolo dispatched hundreds of fighters and bombers to reinforce his units in Cyrenaica and build up a reserve in Tripolitania. Desperate for reinforcements, Pricolo obtained Il Duce's permission in January to withdraw the CAI's 140 surviving warplanes from Belgium. By early February 1941, however, the Regia Aeronautica in Libya had lost nearly 700 aircraft to all causes—some 400 abandoned or burned to prevent capture, 140 destroyed by RAF ground attacks, 100 lost in aerial combat, and the rest to unknown causes. About 70 aircraft remained in operational condition, while nearly 500 heavily damaged planes lay on airfields around Tripoli in unflyable condition. Mussolini, however, instructed Pricolo that Albania took precedence over Libya for fuel, spare parts, and logistics personnel. Given the destruction of most of the North African infrastructure and supply system, those orders led to the paralysis of Italian air operations. The Regia Aeronautica in Libya had virtually ceased to function. Only the late January arrival of 80 Luftwaffe aircraft—60 Ju 87 dive-bombers and 20 Bf 110 long-range fighters—saved the vital port of Tripoli and nearby Castel Benito airport from destruction by RAF bombers. The landing of an additional 60 German warplanes in February–March created the preconditions for the counteroffensive soon launched by the advance units of the Deutsches Afrika Korps.

Yet even under these catastrophic conditions, Pricolo preferred to use whatever forces were not employed in the Balkans or being shot up in North Africa to carry out bombing missions against Malta, rather than to escort convoys to Tripoli or carry out maritime surveillance for the navy. Diminishing numbers of aircraft forced the air chief to slacken attacks on the island in the winter and spring of 1941. Until their departure for the Eastern Front in May, the island's bombardment was increasingly carried out by Luftwaffe warplanes. But in these Malta raids, the Regia Aeronautica lost 60 bombers and fighters, with another 250 damaged beyond repair, between Italy's entry into the war and April 1941. Although it was only one factor in allowing the British carrier strike against Taranto in November 1940, the bombardment of Genoa in February 1941, and the loss of three Italian heavy cruisers at Matapan in March 1941, Pricolo's refusal to provide the navy with the reconnaissance flights it requested contributed significantly to these disasters. Certainly the Regia Aeronautica staggered under the excessive demands placed on it from October 1940 to April 1941. In addition to the losses already cited, it lost 100 aircraft in combat, suffered 230 seriously damaged by accidents, and had 570 put out of action by the enemy in the Greek–Albanian campaign. Flights out of Sardinia in support of naval operations cost the air force about 100 aircraft in the first nine months of the war and about the same number in similar missions from Sicily and the Aegean Islands; several hundred more were seriously damaged. Pricolo admitted that his aircrews found Malta very heavily defended and that targets on the island were extremely difficult to damage. It seems that the aircraft so employed could have been put to better use elsewhere, had Douhet's theories not continued to hold Pricolo and his staff in their grip.

The New High Command and Its Operations, 1941–1943

Aspects of the command system changed radically following Badoglio's dismissal in December 1940, when Mussolini chose him to bear responsibility for the repulse of the Italian invasion of Greece. As successor, Il Duce appointed Ugo Cavallero, an army general with the unusual experience of directing the Pirelli rubber corporation (1920–1925) and the Ansaldo armaments works (1928–1933). The appointment came with Ciano's enthusiastic approval, for he and Cavallero had been closely aligned for several years. Simultaneously, Cavallero assumed command of the shattered Italian forces in Albania and brought the Greek counteroffensive to a halt. After he returned to Rome following the end of the Balkan campaign in May 1941, Cavallero purged the high command of enemies, secured Mussolini's agreement to create a true *Comando Supremo,* and began imposing his new authority and a degree of cooperation on the armed forces. He also brought his industrial expertise to bear on the organization of the Italian wartime economy and recognized the need for the closest possible cooperation with the Germans. To that end, Cavallero appointed Colonel Raffaele Senzadenari, a former air attaché in Berlin, where he had befriended Göring, to serve as his private secretary.

Cavallero and Pricolo soon clashed. The air undersecretary–chief of staff had already attracted Cavallero's dislike for serving as Mussolini's informant about ground operations during the Greek–Albanian campaign. Whereas Pricolo enjoyed Mussolini's confidence, Cavallero possessed the advantage of being Ciano's principal military ally. Since early 1940, Il Duce's son-in-law had attempted to guide the direction of the Regia Aeronautica, but Pricolo's refusal to accept this interference had soured their relations. Friction with the foreign minister combined with Pricolo's difficulties with Cavallero to undermine the air undersecretary's position. Over the summer and fall of 1941, Pricolo increased Cavallero's ire by resisting the efforts of the *Capo di Stato Maggiore Generale* to exercise direct control over the air industry, by expressing anti-German sentiments, and by placing air force interests above those of the overall war effort. In particular, the air chief neglected the aerial escort of convoys to Libya, which were particularly hard-pressed by British attacks in the fall of 1941, in preference for continued raids on Malta. When the new MC.202 fighters became operational in November, Pricolo ignored Cavallero's orders to send them to North Africa, awaiting the installation of sand filters. Regardless of the validity of the technical reasons, the air chief's disobedience of a direct order came at a time when British preparations for the "Crusader" offensive had become clear, and the Italian forces badly needed the new fighter to match the Spitfire. Meanwhile, the destruction of the *Duisburg* convoy bound for Tripoli—a night action on 9 November for which Pricolo was blameless—demanded a scapegoat. The unscrupulous Cavallero and Ciano persuaded Mussolini of the air chief's guilt and demanded Pricolo's removal. Il Duce dismissed him in person on 14 November.

After a frenzied search for a replacement, the air undersecretary and chief of staff posts went to General Rino Corso Fougier. Ciano, who had backed Fougier's selection, also managed to get Colonel Giuseppe Casero appointed as Fougier's ministerial cabinet chief. Ciano and Casero had served together during the Ethiopian war, and after Ciano's appointment as foreign minister in June 1936, Casero had served as liaison between Il Duce's son-in-law and the Regia Aeronautica. Despite obvious opportunities for corrupt behavior, Casero performed his duties with honesty and rigid correctness. Fougier had been professionally wounded by his 1940–1941 command of the poorly performing CAI, and on his return to Italy, he had implored Casero to intervene with Ciano to save him from forced retirement. Though Fougier remained on active duty, the general languished in his new post as air representative on the Italian–French armistice commission; however, the position brought the advantage of frequent contact with Ciano. With the impending transfer of Air Marshal Kesselring from Russia to Italy to become German Mediterranean theater commander and reassume command of Luftflotte II in late November, Fougier seemed a good choice to work with him, as they had collaborated well in the past. Pricolo's dismissal opened the ideal post from which to do so, especially with Casero as Fougier's office manager and go-between with Ciano. In fact, the Fougier–Casero–Kesselring partnership proved highly successful.

By the spring of 1941, the inability of the Italian air industry to replace losses became manifest. Total production from Italy's entry into the war through March 1941 totaled some 3,100 planes. Adding the results of accidents suffered in training or noncombat operations to the numbers cited earlier, the Regia Aeronautica lost about 3,500 aircraft during the same period. However, the modern fighters—the G.50s and MC.200s, finally coming off the assembly lines in early 1941—were delivered without armor or long-range fuel tanks. Until these were installed—a time-consuming process for the hard-pressed air force ground crews—the fighters could not be turned over to operational units. As a result, the first MC.200s were not delivered to Tripolitania and Albania until March 1941.

More significant reinforcement arrived in the form of the 300 German aircraft of X Fliegerkorps sent to Sicily in January 1941. Until the Regia Aeronautica began to recover from its losses in the autumn of 1941, those Luftwaffe warplanes would sustain the bombing campaign against Malta and the escort of convoys to Libya. Likewise, in the May 1941 Axis air attacks on Royal Navy warships evacuating the Crete garrison, XI Fliegerkorps sank three cruisers and five destroyers; Italian bombers from 41 Gruppo on Rhodes sank one destroyer; the Luftwaffe damaged one aircraft carrier, three battleships, five cruisers, and seven destroyers; and the Regia Aeronautica damaged just one cruiser. When X and XI Fliegerkorps withdrew to support the invasion of Russia in mid-1941, however, the Italian air force would undergo another crisis due to its inability to match the superior aircraft of the RAF and replace its losses. It would take the return of an even larger Luftwaffe force to the Mediterranean—Luftflotte II under Kesselring—to restore Axis fortunes in November 1941. Although the Regia Aeronautica would keep fighting until September 1943, enjoying some notable successes in 1942 and always displaying a stoic fortitude, the war had already been lost simply by Mussolini's decision to enter the conflict. However painful for many Italians to admit, then and since, the Italian air force collapsed in early 1941 and survived thereafter only as an appendage of its German ally.

Fougier's command experience alongside the Luftwaffe against the best units of the RAF in 1940–1941 left him profoundly pessimistic when he took over the Regia Aeronautica. He apparently had already decided that the Axis would likely lose the war—an opinion that hardened after American entry into the conflict. Nonetheless, Fougier realized that any chance of victory depended on the closest possible cooperation with the Germans, however difficult or humiliating that might be. By late 1941, the Regia Aeronautica could not operate without the fuel provided by the Germans; nor could the air industry increase production unless it acquired special machine tools from Germany. Fougier needed German high-performance engines for the new Series 5 fighters scheduled for production in 1942. By the spring of that year, the air chief had greatly improved cooperation with the navy. By the fall, he had gotten the MC.205 fighter and the outstanding Cant Z.1018 medium bomber into production, prepared the series manufacture of the Fiat G.55 and the Re.2005, and pushed the air industry toward greater efficiencies. Fougier's efforts allowed the Regia Aeronautica to take a more significant role in the pounding of Malta over the winter of 1941–1942 and into the spring, although II Fliegerkorps inflicted most of the damage. During the murderous assault on the "Pedestal" convoy to Malta in mid-August 1942, more than half of the nearly 800 Axis aircraft participating were Italian; the arrival of the vital cargoes can be blamed on the failings of the Italian navy leadership, not those of the air force. Throughout the battles in Egypt in the summer and fall of 1942, two-thirds of the heavily outnumbered Axis aircraft supporting Rommel's forces were Italian. They performed heroically while being virtually annihilated during the final battle over El Alamein and covering the retreat into Tunisia. These Regia Aeronautica accomplishments pleased Cavallero, although Fougier had privately come to despise the *Capo di Stato Maggiore* for his groveling before the Germans and for his falsely optimistic reports to Mussolini. By the end of 1942, Fougier could no longer hide his disdain for Cavallero. At the same time, the growing friction between the Axis allies in the months preceding Il Duce's downfall turned Fougier's determination to work smoothly with Kesselring and other German leaders from an asset into a liability. Furthermore, Ciano's dismissal as foreign minister and Cavallero's removal as head of *Comando Supremo* in February 1943 deprived the air undersecretary of his erstwhile patrons. Only Mussolini's personal insistence on Fougier's retention—the general's performance had gained Il Duce's approval—saved him for a time. During those months, the air force expended its rapidly diminishing resources, suffering crippling losses to the British and Americans while maintaining the Axis air bridge to Tunisia in November 1942–May 1943. Most of the remaining strength of the Regia Aeronautica went into the pre- and post-invasion defense of Sicilian airspace, a doomed struggle symbolized by the loss of two of the three leading Italian aces in separate incidents over the island on 5 July 1943. The air force's failure to put up more than a token defense of Rome during the massive American bombing attack of 19 July and the indiscipline of ground personnel at airfields around the city that day infuriated the king. A week later, following the overthrow of Mussolini, Fougier was replaced by General Renato Sandalli, the deputy air chief of staff for armaments.

Although Badoglio took Mussolini's place as head of the government, the elderly marshal had no desire to assume the service portfolios as well. The new prime minister had vague but positive memories of Sandalli from the Ethiopian war, when he had commanded a bomber unit, and air intelligence indicated the general's political reliability. In any case, there appeared to be no suitable alternative, so the new prime minister promoted Sandalli to be both air minister and chief of staff. But by late July 1943, the Regia Aeronautica had been virtually shot from the skies, and Sandalli presided over a shattered force. Meanwhile, the Badoglio government secretly opened negotiations with the Allies to change sides in the war. Unfortunately, Badoglio's limited trust of Sandalli did not extend to informing him of such clandestine maneuvers or ordering him to make appropriate preparations. Warned only a few days before the armistice, but given no directions, the air force leadership operated in confusion. Should it deploy its aircraft to defend Rome, flee to Allied-controlled airfields in the south, or protect the Italian fleet as it sought to escape capture by the Germans? Sandalli neither received nor gave instructions to settle these questions. As a result, the Regia Aeronautica collapsed in disorder from 8 to 10 September, with some units joining the Germans, others being taken prisoner, and a few reaching areas under Allied control. Sandalli and his staff managed to join the Badoglio government, which had taken refuge in Brindisi.

In September1943, the remnants of the Italian air force were divided in two: a Regia Aeronautica of about 200 aircraft evacuated to or rebuilt in the south, and the new Aeronautica Nazionale Repubblicana (ANR), created with Luftwaffe assistance. Officially, the ANR served Mussolini's puppet Italian Social Republic, which provided camouflage for German control of central and northern Italy. In January 1944 the ANR began flying some 100 of the roughly 1,200 operational Italian aircraft seized by the Germans. The ANR consisted of five understrength groups: two of fighters, one each of torpedo bombers (using the old but reliable SM 79) and transports, plus an aircraft ferry group. The transports served only on the Eastern Front. The ferry group transferred to the Reich most of the aircraft produced by Italian plants working for the German war effort. Initially, the ANR fighter groups also received Fiat and Macchi-Castoldi Series 5 fighters. But in the summer of 1944, after successfully resisting incorporation into the Luftwaffe and burning their G.55s and MC.205s, they were reequipped with Bf 109Gs. Using both types of fighters, six ANR fighter pilots scored five or more victories against Allied bombers.

The Regia Aeronautica returned to action immediately after the September 1943 armistice, flying transport and supply missions for Italian units cut off in the Balkans. It was amazingly successful at keeping its Italian aircraft flying and also began receiving Allied-supplied Spitfire and P-39 Aircobra fighters and Baltimore medium bombers. By late 1944, the Regia Aeronautica had been rebuilt to a force of nearly 300 warplanes and 600 pilots. Throughout the last year and a half of the conflict, its combat missions consisted largely of ground attack, maritime interdiction, and transport flights in and out of enemy-held territory, although Regia Aeronautica pilots also downed a dozen German aircraft. The two Italian air forces avoided air-to-air combat with each other.

Both Italian air services performed with valor and dedication. But considering the cause for which they flew, it is difficult to accept ANR veterans' explanations that they had fought only to defend northern Italian cities or "for the honor of Italy." In contrast, the Regia Aeronautica contributed more than is generally realized for the liberation of Italy and the Balkans. In 1946 it was renamed the Aeronautica Militare Italiana of the newborn Italian Republic.

Conclusion

The defeat of the Regia Aeronautica sprang directly from incoherent air-war doctrine, the air industry's inability to adopt mass production techniques, and Italian technological backwardness. An effective use of the aircraft then available could have granted the Regia Aeronautica significant victories in 1940–1941. Substantially increased production of warplanes in 1941–1942 could have buffered the difficult transition to an effective air strategy. Sufficient advanced aircraft in 1942–1943 could have delayed Italian defeat. But all three weaknesses led to the effective collapse of the Italian air force in early 1941. Only German assistance allowed the Regia Aeronautica to continue fighting until it was obliterated by well-directed and overwhelming numbers of modern Allied aircraft in 1943. But the fundamental reasons for the downfall of the Regia Aeronautica lay in Mussolini's disastrous strategic leadership and the incompetent, corrupt, and feckless Fascist regime he had created to rule Italy.

Suggestions for Further Research

Valle and Pricolo each produced two memoirs, but they present apologetic viewpoints. The Italian air force historical office also published a short, rather exculpatory biography of Valle. Fougier published no autobiography. Scholarly biographies of these three air force leaders would offer many insights about the Regia Aeronautica from 1923 to 1943. In addition, Pricolo kept a diary, an edited version of which might prove invaluable.

Valerio Castronovo wrote an excellent biography of Giovanni Agnelli, the founder and longtime head of Fiat, but it touches on airframe and air engine design, contracts, and production only lightly. The lives of several other air force and air industry figures would make excellent subjects for biographies to help expand knowledge of the history of the Regia Aeronautica, including the industrialists Rinaldo Piaggio and Giovanni Battista "Gianni" Caproni (studies of their influence on Douhet and Mussolini would explain much); generals whose careers spanned the first decades of Italian airpower, such as Sandalli, Alessandro Guidoni, Mario Bernasconi, Eraldo Ilari, Gennaro Tedeschini Lalli, Mario Aimone Cat, Ferruccio Ranza, Felice Porro, Silvio Scaroni,

Pietro Pinna, and Ruggero Bonomi; and the aeronautical engineers Secondo Campini, Giuseppe Gabrielli, Antonio Ambrosini, Mario Castoldi, Alessandro Marchetti, Celestino Rosatelli, and Filippo Zappata (accounts of their thwarted brilliance would describe the spectacular rise and swift collapse of Italian airpower). Gabrielli did write a short, unrevealing autobiography, and Caproni and Scaroni both kept diaries, which have been withheld from publication by their heirs. Given widespread diary keeping and memoir writing among well-educated Italians at the time, many other such accounts may exist. Those by generals, aircraft designers, and industrialists could provide valuable information on relations among the air force, air industry, and Fascist regime.

Official histories of the Regia Aeronautica have been published in Italy, but a balanced, scholarly account of the Italian air force during the Second World War, based on Italian and foreign documentation and more recent secondary sources, is yet to be written.

An investigation of the financial links among industrialists, air force commanders, and Fascist leaders would explain much about the fate of the Regia Aeronautica. Although Ciano's diaries have been published and several biographies have been written, his air force career and his influence on the Regia Aeronautica remain to be explored. Research for a history of SIA could prove difficult, given the reported burning (and certain disappearance) of its files in September 1943. But an account drawn from other intelligence sources would help explicate what Mussolini and his air generals knew and how much they chose to ignore about their aerial enemies during World War II.

The U.S. National Archives holds a mass of captured military documentation in Italian (microfilm series T-821) and German (microfilm series T-78) on the Regia Aeronautica during World War II. Guides are available for both series, and any reel can be purchased. Thus, considerable primary research on the Italian air force from 1940 to 1943 can be carried out without travel.

Recommended Reading

There are few reliable publications on the Regia Aeronautica in English, and despite a large number of books on the topic, *reliable* histories are nearly as rare in Italian. English-language books on the Italian army and navy have begun to appear in increasing numbers over the past twenty years. The Italian air force, however, remains far less studied by historians on both sides of the Atlantic. Thus, in the unlikely event of the translation of all books on the Regia Aeronautica in World War II, relatively few of high value would become available in English. This is not meant to disparage the works cited but merely to explain their small number and the lacunae among them.

Walter J. Boyne, ed., *Air Warfare: An International Encyclopedia,* 2 vols. (Santa Barbara, Calif.: ABC-CLIO, 2002), contains several solid articles on Italian subjects. Those by Gregory Alegi on Regia Aeronautica history, its role in the Ethiopian war and the Second World War, and Italian air personalities, aircraft, and manufacturers are noteworthy. John Ellis, *World War II: The Encyclopedia of Facts and Figures* (n.p.: Military Book Club, 1995), provides information on command structure, order of battle, strength, losses, and aircraft specifications of the Italian air force, as well as aircraft production figures.

Although Italo Balbo died less than three weeks after Italy's entry into World War II, his influence on the air force was significant. Administering and inspiring the Regia Aeronautica constitute only part of Balbo's life story, but a basic understanding of air force policies and politics before and during the Second World War can be acquired from Claudio G. Segrè, *Italo Balbo: A Fascist Life* (Berkeley: University of California Press, 1987). Although he passed away in 1930, Giulio Douhet left an equally important imprint on the air force of 1940–1943. Much of Douhet's mature airpower theory is contained in the translation of the second edition (1927) of his *Command of the Air* (Washington, D.C.: Office of Air Force History, 1983). Segrè was in the process of writing a biography of Douhet at the time of his premature death; despite this loss, Segrè's basic views on the theorist can be gained from his "Douhet in Italy: Prophet without Honor?" *Aerospace Historian* (June 1979). A partial substitute for a Douhet biography in English is provided by the dissertation of another scholar who suffered an early demise: Frank J. Cappelluti, "The Life and Thought of Giulio Douhet" (Rutgers University, 1967). Since Galeazzo Ciano was an air force reserve bomber pilot, as well as Il Duce's son-in-law and foreign minister (1936–1943), he had great interest in Regia Aeronautica affairs. A thorough knowledge of the Fascist regime helps clarify Ciano's sometimes cryptic journal remarks on personalities and power struggles. But even without that background, Ciano's *Diary 1937–1943,* ed. Renzo De Felice and Stanislao G. Pugliese (New York: Enigma Books, 2002), provides considerable information on the air force leadership during the Spanish Civil War, the events prior to World War II, and the conflict itself. Ray Moseley, *Mussolini's Shadow: The Double Life of Count Galeazzo Ciano* (New Haven, Conn.: Yale University Press, 1999), offers a fuller understanding of Ciano's (and Mussolini's) influential attitudes about the air force and airpower. Several dozen articles by myself and MacGregor Knox on other air force personalities and various Regia Aeronautica activities throughout the Fascist period can be found in Philip V. Cannistraro, ed., *Historical Dictionary of Fascist Italy* (Westport, Conn.: Greenwood Press, 1982).

The place of airpower in the armed forces, the air force, and Mussolini's strategic thinking—quite separate concepts—is not addressed comprehensively in English. Nonetheless, a good deal can be learned from works touching on Italian military policy and preparations before World War II, as well as on Regia Aeronautica involvement in the Ethiopian and Spanish civil wars and in the Second World War itself. A brief survey of Regia Aeronautica strategic orientation between the world wars is available in my "The Italian Armed Forces, 1918–40," in *Military Effectiveness,* 3 vols., ed. Allan R. Millett and Williamson Murray (Boston: Allen and Unwin, 1988), vol. 2, *The Interwar Period.* My dissertation, "A Thirst for Glory: Mussolini, the Italian Armed Forces and the Fascist Regime, 1922–1936" (Columbia University, 1984), and George W. Baer, *Test Case: Italy, Ethiopia and the League of Nations* (Palo Alto, Calif.: Stanford University Press, 1976), both offer solid accounts of air force activities in the Ethiopian war. John F. Coverdale, *Italian Intervention in the Spanish Civil War* (Princeton, N.J.: Princeton University Press, 1975), and my article, "Fascist Italy's Military Involvement in the Spanish Civil War," *Journal of Military History* (October 1995), do the same for the 1936–1939 conflict in Spain, although Coverdale's book concentrates on the first year of the fighting. Three sound books that focus on other topics nevertheless give general accounts of the evolution and employment of aspects of Italian airpower (1935–1943) in larger contexts, and each draws on a wealth of primary and secondary Italian sources: Robert Mallett, *The Italian Navy and Fascist Expansionism 1935–1940* (London and Portland, Ore.: Frank Cass, 1998); Reynolds M. Salerno, *Vital Crossroads: Mediterranean Origins of the Second World War, 1935–1940* (Ithaca, N.Y.: Cornell University Press, 2002); Jack Greene and Alessandro Massignani, *The Naval War in the Mediterranean 1940–1943* (London: Chatham Publishing; Rockville Centre, N.Y.: Sarpedon Publishers, 1998). Greene and Massignani also authored *Rommel's North African Campaign, September 1940–November 1942* (Conshohocken, Pa.: Combined Books, 1994). Despite its misleading title, the book provides a detailed history of Italian forces in the North Africa campaign through El Alamein. The authors concentrate on ground actions but also describe the Italian strategic and operational context in which the Regia Aeronautica fought its North African war. MacGregor Knox has written several outstanding books and articles on Italy in World War II. His *Mussolini Unleashed, 1939–1941: Politics and Strategy in Fascist Italy's Last War* (Cambridge: Cambridge University Press, 1982); "The Italian Armed Forces, 1940–43," in Millett and Murray, *Military Effectiveness,* vol. 3; and *Hitler's Italian Allies: Royal Armed Forces, Fascist Regime, and the War of 1940–1943* (Cambridge: Cambridge University Press, 2000) all provide much information on and analysis of Italian air strategy in the Second World War. Though written from a German perspective, the English-language translation of the Research Institute for Military History's multiauthor series, *Germany and the Second World War* (Oxford: Oxford University Press, 1990–), offers considerable information on the Italian war effort. Volume 3, *The Mediterranean, South-east Europe and North Africa, 1939–1941: From Italy's Declaration of Non-belligerence to the Entry of the United*

States into the War, uses many Italian secondary sources and covers the activities of the Regia Aeronautica and its partnership with the Luftwaffe from June 1940 to December 1941. Finally, for the serious researcher or patient reader, much information about Italian air policy (1933–1941), also seen through German eyes, can be gleaned from the nineteen volumes of *Documents on German Foreign Policy 1918–1945,* Series C and D (Washington, D.C.: GPO; London: HMSO, 1949–1983). Volume 3 of Series D on the war in Spain and the later volumes of the same series help explain German attitudes about the Regia Aeronautica.

Two books give extensive information on the development and performance of Italian military aircraft types employed in World War II. Jonathan Thomson, *Italian Civil and Military Aircraft 1930–45* (Fallbrook, Calif.: Aero Publishers, 1963) remains unrivaled for accurate detail. The Italian origins of Enzo Angelucci, *The Rand McNally Encyclopedia of Military Aircraft 1914–1980* (Chicago: Rand McNally, 1981), explains its extensive coverage of Regia Aeronautica warplanes. Both works provide many excellent photos and drawings, as well as valuable charts and appendixes. Though less detailed, another book translated from Italian—Enzo Angelucci and Paolo Matricardi, *Complete Book of World War II Combat Aircraft* (Chicago: Rand McNally, 1978)—presents both technical information on and large, detailed paintings of the major Regia Aeronautica fighters and bombers in use from 1940 to 1943. Geoffrey D. M. Block, *Allied Aircraft versus Axis Aircraft* (Old Greenwich, Conn.: WE Inc., 1968), presents firsthand evaluations of some Italian warplanes studied in North Africa and Sicily during World War II and discusses their merits relative to their British counterparts.

For a balanced overview of the Italian air force (1940–1943), see Hans Werner Neulen, *In the Skies of Europe: Air Forces Allied to the Luftwaffe, 1939–1945* (Ramsbury, U.K.: Crowood Press, 2000). Chris Dunning, *Courage Alone: The Italian Air Force, 1940–1943* (Aldershot, U.K.: HIKOKI Publications, 1998), offers a great deal of information that is otherwise inaccessible to nonreaders of Italian. Dunning provides accurate details on command structure, unit histories, orders of battle, operations on all fronts, aircrew training, aces' scores, aircraft data, weapons, equipment, markings, camouflage, and the strengths and weakness of the service, as well as a multilingual bibliography. Christopher Shores and his coauthors have written detailed operational chronologies of the campaigns involving Axis and Commonwealth air forces: Shores and Hans Ring, *Fighters over the Desert: The Air Battles in the Western Desert, June 1940 to December 1942* (London: Spearman, 1969); Shores, Ring, and William N. Hess, *Fighters over Tunisia* (London: Spearman, 1975); Shores, Brian Cull, and Nicola Malizia, *Air War for Yugoslavia, Greece and Crete, 1940–41* (London: Squadron Signal Publications, 1987); Shores, Cull, and Malizia, *Malta: The Hurricane Years, 1940–41* (London: Grub Street, 1987); Shores, Cull, and Malizia, *Malta: The Spitfire Year, 1942* (London: Grub Street, 1991); Shores, *Dust Clouds in the Middle East: The Air War for East Africa, Syria, Iran and Madagascar* (London: Grub Street, 1996). Shores has also assembled two fine photo collections: *Pictorial History of the Mediterranean Air War* (London: Ian Allen, 1972) and *Regia Aeronautica: A Pictorial History of the Italian Air Force, 1940–1943* (Warren, Mich.: Squadron Signal Publications, 1976). Two Italian air campaigns are described in Thomas Potts, "L'Operazione Cinzano," *Aerospace Historian* (March 1981), and Anthony Rogers, *Battle over Malta: Aircraft Losses and Crash Sites, 1940–42* (Stroud, U.K.: Sutton, 2000).

Little in English has appeared on the Italian economy, and even less on the aircraft industry. Still, Vera Zamagni, *The Economic History of Italy, 1860–1990* (Oxford: Oxford University Press, 1993), provides a concise introduction. Two dissertations, one in book form, offer aviation-specific information: Angela Raspin, *The Italian War Economy, 1940–1943: With Particular Reference to Italian Relations with Germany* (New York: Garland Publishing, 1986); Cristiano Andrea Ristuccia, "The Italian Economy under Fascism: 1934–1943: The Rearmament Paradox" (Oxford University, 1998). On Italian aircraft engines, consult Herschel H. Smith, *A History of Aircraft Piston Engines: From the Manly Balzer to the Continental Tiara* (Manhattan, Kans.: Sunflower University Press, 1986).

CHAPTER SIX

The Imperial Japanese Air Forces

Osamu Tagaya

Following Japan's defeat in World War II, former vice admiral Torao Kuwabara, one of the leading pioneers of aviation in the old imperial navy, declared that Japan's mistake had been starting the Pacific war too soon. By this, he meant that Japan had initiated armed conflict with the Western powers before its technological and industrial capabilities had matured sufficiently to allow it to confront them successfully in a major war.

Certainly, by any measure of economic activity, the disparity in industrial strength between Japan and its Allied opponents, particularly the United States, was of such magnitude that, in retrospect, it is difficult to envision how Japan ever hoped to attain victory in that global conflict. Closely related to industrial capacity is manpower. Japanese veterans of the Second World War hold vivid memories of overwhelming enemy numbers, and it is widely believed among the Japanese people that they were defeated by the quantitative superiority of their foes. This belief may be some consolation to those looking to salvage some redeeming aspect of their sacrifice from the wreckage of national defeat; implicit in this view is the notion that, but for the enemy's superior numbers, Japan's own fighting qualities would have shone through brightly. Attributing their defeat to enemy quantity also conveniently avoids a probing analysis of the quality of their conduct of the war. The broad question of whether Japan's defeat was inevitable is beyond the scope of this chapter, but it is worth noting that, like its Axis partner Nazi Germany, the defeat of imperial Japan was total. Many factors underlying the defeat of Japan's air forces involve areas beyond the mere conduct of air operations or the technical capabilities of the aviation industry. In dealing with Japan, therefore, we must view its aerial defeat against the backdrop of a wider canvas.

Roots of Defeat

There is no denying that the technological, industrial, and numerical superiority of Japan's enemies played a major part in its defeat. But the roots of Japan's ultimate fate in World War II run much deeper. By the middle of the 1930s, Japan found itself saddled with a fundamentally dysfunctional policy of national defense. The structure of government formalized by the Meiji Constitution of 1889 had subordinated the army directly to the imperial throne, giving it equality with, rather than making it subordinate to, the civilian arm of government. The navy was accorded similar independence in 1893. In the decades following Japan's victory in the Russo-Japanese War of 1904–1905, these two branches of the imperial armed forces came to espouse increasingly divergent views on national security and the identity of major external threats to the nation. The army continued to view Russia (and its successor state, the Soviet Union) as its main potential enemy—a view that only increased with Japan's ever-growing involvement on the Asian mainland. The navy, in contrast, came to see the United States as the main hypothetical enemy, with the vast expanse of the western Pacific Ocean as the main battlefront in any future conflict. In the absence of oversight by the civilian government, which might have provided a unifying influence and allowed the nation to formulate a cohesive policy of national defense, the army and navy, driven by their own partisan ambitions, each pursued a separate course and devised strategy, developed weaponry, and trained personnel to suit their own narrow interests. In a nation already challenged by its inferiority in manpower and industrial capacity in comparison to its potential enemies, this state of affairs merely aggravated competition for scarce national resources. The waste and inefficiency created by duplication of effort and lack of uniformity in procedures and equipment between the army and navy, down to the most basic levels, were major contributing factors in Japan's defeat. In the field of aviation specifically, engines, armaments, radio equipment and voltage, spare parts, and grades of fuel all differed between the air forces of the two rival services. On two separate occasions during the interwar period, the army attempted to establish an independent air force, but these efforts were opposed by the navy and came to naught. Belated efforts by the Ministry of War Production (from 1943 onward) to introduce uniform standards and boost productivity through the strategic allocation of resources came too late and had only limited success. Given the long history and systemic nature of the problem, it is doubtful that any effort at rationalization launched after the start of the Pacific war would have succeeded in resolving this vital issue. In addition, it seems unlikely that a sufficient sense of urgency could have been mustered before the start of the conflict to address this problem in any meaningful fashion. If, however, a more cohesive policy regarding national defense had been implemented during the early decades of the twentieth century, it is likely that Japan's air forces would have been better prepared to fight a long war with the formidable adversaries they faced after 1941.

Army Shortcomings in the Air

The army's air force, in particular, found itself poorly prepared to conduct the war it was eventually called on to fight. Having trained and equipped to fight a limited tactical war on land against the Soviet Union in the Far East, the Japanese Army Air Force (JAAF) lacked personnel capable of long-distance over-water navigation and possessed short-range aircraft that proved to be of limited use in the archipelagoes of Southeast Asia and the Pacific. At the start of the Pacific war, long-range penetration raids against major targets in southern Luzon in the Philippines had to be assigned to the navy's Type 1 and Type 96 land-based, twin-engine attack aircraft (G4M1 "Betty" and G3M2 "Nell"), which had the necessary range to reach these targets. The twin-engine but shorter-range Type 97 heavy bombers (Ki 21 "Sally") and Type 99 light bombers (Ki 48 "Lily") of the army had to content themselves with lesser targets in central and northern Luzon, closer to base.

The fact that the navy's air service had produced its own long-range bombers reflects the peculiar duplication of mission roles that had developed within the Japanese air forces. The Imperial Japanese Navy (IJN) had conceived its land-based attack (*rikko*) category of aircraft in 1932 as a means of offsetting the lower warship tonnage ratios imposed on it in comparison to the British and American navies by the London Naval Treaty of 1930. Aircraft were excluded from the treaty limitations, and the navy envisioned these planes supporting the fleet by attacking enemy warships with torpedoes or bombs, flying from island bases and operating over long ocean distances. In some respects, the development of the *rikko* addressed the same strategic challenges that led to the development of the Boeing B-17 in the United States. In the latter case, the U.S. Army Air Corps had sought, in part, to justify its development of a four-engine heavy bomber by promoting it as a means of providing long-distance protection for America's shores, thereby precipitating a turf battle with the U.S. Navy. But whatever the justification used to argue its position, the army's desire to develop a long-range heavy bomber was grounded firmly in strategic concepts of airpower. These concepts had been expounded by the evangelical rhetoric of Billy Mitchell and would eventually be put into practice by a younger generation of army officers, such as Henry H. Arnold, Carl Spaatz, and Ira Eaker, many of whom had firsthand knowledge of military aviation through their service as airmen in the First World War.

A similar chorus of air advocacy was sadly lacking in the Imperial Japanese Army. The Japanese had been among the first to field aircraft in combat during the Great War, when both the army and the navy had dispatched a number of Maurice Farman biplanes and other types to northeast China during the siege of the German fortress of Tsingtao.

These machines confirmed the value of the airplane in reconnaissance work, undertook some rudimentary bombing missions, and made the first faltering attempts at air combat against the lone Rumpler Taube flown by the Germans. But the fall of Tsingtao in November 1914 marked the end of Japanese participation in the land battles of World War I. The quantum leaps in aviation, both technical and operational, that occurred above the killing fields of France and Flanders largely passed them by. Unlike other major participants in World War I, the Japanese missed the opportunity to build a cadre of young aviators with direct combat experience. They were not without their share of visionaries and air advocates, but the Japanese army lacked a sufficiently large core group of officers with firsthand experience who could champion an independent air doctrine and who could, two decades later, be in a position to provide strong and experienced leadership for its air corps. This would ultimately have grave consequences for Japanese army aviation.

To be sure, there were those in the army's senior ranks who recognized the value of aviation early on. In 1920 Lieutenant General Ikutaro Inoue, head of the army's Department of Aviation, lobbied strongly for the creation of an independent air force as a third branch of the imperial armed forces. But the majority of the members of a joint army–navy committee established to study the matter felt that such a bold move was premature, especially since aviation had yet to attain independent corps status within the army itself. In addition, the navy refused to endorse the proposal, believing that an independent air force would be oriented toward land-based aviation and would not be sensitive to the special needs of the navy. The IJN duly noted that in Great Britain, under the recently established Royal Air Force (RAF), naval aviation appeared to receive low priority.

Discouragement also came from another quarter. Although Japan had taken its seat at the Versailles peace conference at the end of World War I as one of the Great Powers—a remarkable achievement for a nation that only a generation before had been a preindustrial, feudal state—the psychology of a developing nation instinctively looking to the West for leadership and guidance still prevailed. Realizing how badly

Japan had fallen behind the West in the field of aviation during the First World War, the Japanese army had arranged for an aviation mission from France, the nation with the largest air force at the end of the war, to help modernize Japanese military aviation. The mission, composed of some sixty individuals headed by Colonel Jean Faure, arrived in Japan in January 1919, bringing with it a selection of the latest French military aircraft. In retrospect, Faure was probably not an ideal choice to head the mission. When Japanese army leaders asked his opinion about giving aviation greater independence, Faure advised against it. An artillery officer by background, he counseled against giving aviators separate corps status even within the army. He pointed out that career advancement for officers would be a serious problem if aviation were given corps status, since it constituted only a tiny portion of the army's organization. He also believed that effective support of ground troops by aviators required their thorough understanding of traditional ground-based roles. Although he expounded forcefully on the value of aviation, memoranda to his Japanese hosts make it clear that Faure saw aviation's value exclusively in terms of tactical support for ground forces. The influence of such advice, as well as the inherently conservative views of the army leadership, would hold back the early development of aviation in the Japanese army and lock it into a tactical bias for many years.

Reflecting this slow progress, the mission roles assigned to army air units remained primarily reconnaissance and pursuit throughout the 1920s, with bombardment trailing a distant third. The army's first bomber regiment, the 7th Hiko Rentai, was not activated until March 1925, and it remained the only unit specifically assigned to a bombardment role for the rest of the decade. In fact, its buildup was extremely slow, and its full establishment of four squadrons *(chutai)* was not completed until January 1930.

For a time, when the aftermath of the Bolshevik Revolution brought a period of diminished threat from its traditional Russian enemy, the Japanese army turned its attention to the possibility of war with the United States and addressed the issue of how aviation might be employed in an invasion of the Philippine Islands. At the beginning of

1928, Lieutenant General Inoue proposed the development of a superheavy bomber capable of bombing Manila from bases on Taiwan (Formosa). Based on specifications issued in March of that year, the Mitsubishi Aircraft Company duly arranged for the licensed production of a modified Junkers K-51, the bomber derivative of the German four-engine Junkers G-38 transport. The aircraft was given the army short-code designation Ki 20, and the first prototype made its maiden flight in October 1931 after more than three and a half years of development. The giant machine was eventually adopted for service use as the Type 92 heavy bomber in August 1933, but by then, the pace of aeronautical progress had made this aircraft obsolete. Production was canceled in 1935 after only six planes had been completed. The project had given the Japanese army and Mitsubishi valuable experience in the design, construction, and operation of very large aircraft, but budgetary constraints, as well as a resurgent Russian threat on the Asian mainland, curtailed interest in this type of aircraft.

Another challenge facing the army in a possible invasion of the Philippines was the problem of how to deploy aircraft rapidly at the invasion beachhead following a landing. Traditional methods of airfield construction using manual labor were deemed too slow. Around 1932, therefore, the army began working on an ingenious solution to the problem, applying naval methods for short takeoffs and landings to land bases. Code-named the "Ke-go Mechanism," this consisted of a catapult launching machine for takeoffs and arresting wires laid on the ground to stop aircraft after a short landing roll. Like the development of long-range heavy bombers, however, this intriguing project was discontinued when the army refocused on a land war in Asia. Had the Japanese army maintained its concentration on a possible war with the United States during the 1930s, it might have continued to improve its ability to project airpower over long distances and to support landing operations against hostile shores. Such efforts certainly would have benefited the JAAF greatly during the Pacific war. But events of the Japanese army's own making would overtake and curtail such developments.

The takeover of Manchuria by the Japanese Kwantung army during 1931–1933 resulted in an extended land frontier with the Soviet Union. Meanwhile, Stalin's first Five-Year Plan, launched in 1928, was achieving great success, allowing rapid industrialization of the Soviet Union and a strengthening of its armed forces. Aggressive moves by the Japanese army in Manchuria only served to heighten Soviet concern for its Far Eastern borders, and it began a buildup of forces in the region. In turn, this buildup of forces and the Soviets' development of the four-engine Tupolev TB-3 bomber caused grave concern among the Japanese. News of the TB-3's deployment to the Soviet Far East, particularly to bases around Vladivostok, led to the realization that the Japanese Home Islands were no longer immune from air attack. By the mid-1930s, therefore, the Japanese army finally began to think seriously about the use of airpower on a more strategic level. The role of aerial bombardment took on new importance, and army leaders sought to create a first-strike capability against the growing Soviet air forces in the region. But initially, the army's newfound interest in the wider implications of airpower appears to have been limited to early destruction of the enemy's air units at their bases, to prevent them from exerting undue influence on the battlefield. There is little mention in documents of the period about attacking rear areas or interdicting lines of communication to isolate the battlefield.

Foreign developments in aviation, rather than their own convictions, continued to sway the army leadership as the 1930s progressed. A 1935 fact-finding mission on the state of the world's air forces dispatched by the Japanese army to the major nations of Europe and the United States returned with the recommendation that greater independence be granted to the army's air arm. A separate mission sent specifically to observe the state of military aviation in Germany the following year returned with even stronger recommendations, calling for an independent air force; the fact finders had been tremendously impressed by the rapid expansion and technical progress achieved by the Luftwaffe. Thus, in 1936 the army again proposed the establishment of an independent air force. Once more, the navy turned it down, raising essentially the same objections it had cited sixteen years earlier. The navy was now also armed with the specific example of the British Fleet Air Arm, which had suffered serious retardation in its development under RAF control. The IJN was, by then, well on its way to establishing a formidable carrier-borne air fleet and had begun to build its own force of long-range, land-based bombers. There was little chance of the navy ceding control over its air arm, and from a purely naval point of view, the IJN was no doubt correct in questioning the value of doing so.

The army finally decided, on its own, to start a major expansion of its air arm and to give it greater autonomy within the army's structure. In April 1937, destruction of an enemy's air force officially became the primary mission of army aviation. This did not please traditionalists in the Japanese army, who continued to see the infantry as the most important element of the military. In July of that year, Japan was drawn into the quagmire of a general undeclared war in China. If anything, the experience of the JAAF on the battlefield in north China and in the conflict with Soviet forces in 1939 along the Manchurian–Mongolian border at Nomonhan was largely tactical in nature. In 1940, official army air doctrine actually took a step backward. While maintaining its emphasis on the destruction of enemy air forces, it was now declared that the main objective of such efforts should be the indirect support of ground forces. As a consequence of such attitudes, the JAAF's offensive power remained essentially tactical during the Pacific war that followed. Despite a number of design projects initiated during those years, the JAAF never completed a flying prototype of a four-engine bomber, let alone introduced it into service.

It is questionable that the mere possession of an operational four-engine bomber would have given the JAAF a significant strategic capability. Aviation in Italy and the Soviet Union during World War II comes to mind in this regard. But greater awareness of the full potential of airpower earlier in its history and a vigorous pursuit of appropriate doctrine in an attempt to realize that potential would have made the JAAF a more potent adversary in the Pacific war. A greater number of air-minded senior army officers and a broader base of junior officers with aviation experience in the early, formative years of its air corps may have led to the establishment and pursuit of such doctrine. Furthermore, a stronger cadre of such men would have allowed more effective leadership of the JAAF at higher levels of command during the Pacific war. Instead, Japanese army aviation during the interwar period was characterized by a general lack of initiative, constrained by the hidebound attitudes of senior officers, and reactionary in responding to international developments. Confronted with a shortage of officers with solid aviation experience during the rapid expansion in the late 1930s and the Pacific war years, the JAAF was forced to draw on personnel from other branches of the service with little or no experience in aviation. This often resulted in poor leadership and unimaginative staff work, giving rise to operations that were questionable in their effectiveness and all too predictable and conventional in nature.

Navy Contributions to Aviation Failure

It is difficult to level similar criticism against the development of Japanese naval aviation. By the end of the First World War, the imperial navy, like its army counterpart, seriously lagged behind the West in the development of its air arm. Following the army's example, the navy prevailed on its traditional Western mentor, Great Britain, to provide assistance in a thorough overhaul and modernization of its air service. The British responded by sending a thirty-three-man mission in 1921 headed by Sir William Francis-Forbes Sempill, a Scottish aristocrat experienced in naval air operations and the design and testing of naval aircraft. The navy's cadre of early aviators was no larger than that of the army, but because of its basic need to operate modern warships in the normal course of duty, the navy enjoyed a greater understanding of and appreciation for technology and its innovations at an institutional level. It provided a more liberal climate in which such air-minded officers as Isoroku Yamamoto, Shigeyoshi Inoue, and Takijiro Ohnishi would eventually rise to flag rank and exert control over the levers of naval power at the highest levels.

The IJN achieved a number of pioneering milestones in naval aviation, including completion in 1922 of the world's first purpose-built aircraft carrier, the *Hosho.* With the Type 96 carrier fighter (A5M "Claude"), it achieved operational deployment in 1937 of the world's first all-metal monoplane fighter designed and equipped for carrier operations. By 1940, its fleet exercises incorporated massed, coordinated attacks by torpedo bombers, dive-bombers, and fighters from several aircraft carriers operating in consort, a skill no other navy in the world then possessed. On the eve of its attack on Pearl Harbor, the IJN possessed ten aircraft carriers, more than any other navy in the world, and wielded the world's most formidable naval air arm. Its Type 0 carrier fighter (A6M "Zeke"), the famous "Zero," was superior in performance not only to its counterparts deployed by other navies but also to its land-based contemporaries. In the development of its land-based attack category of aircraft, the IJN was unique among the world's navies in fielding what was essentially its nation's strategic bombing capability. In other nations, this was the preserve of land air forces, but the IJN usurped this function from the Japanese army, which was myopically focused on a short-range tactical war on the Asian mainland. The sinking of the Royal Navy's battleship HMS *Prince of Wales* and battle cruiser HMS *Repulse* by these aircraft off the Malayan coast on the third day of the Pacific war marked the first time in history that aircraft alone sank capital ships in full combat maneuver at sea, thereby dethroning the battleship as the final arbiter of naval warfare. Meanwhile, in its dramatic raid on Pearl Harbor and its subsequent six-month rampage across a third of the globe, the IJN became the first navy in the world to operate aircraft carriers en masse, ushering in the age of the aircraft carrier as the dominant power at sea.

Yet, for all its power and pioneering achievements, the IJN air service harbored significant flaws that contributed to its ultimate defeat. Although the IJN had pioneered the development of massed carrier airpower, its formal fleet organization continued to reflect a structure centered around its battleships until the formation of the carrier-centered Third Fleet in July 1942. It never developed a fully fledged carrier task force organization incorporating a fleet train capable of supporting carrier operations on a sustained basis. Its U.S. Navy opponent would eventually do so during the second half of the Pacific war. Even as it was adding the formidable fleet carriers *Shokaku* and *Zuikaku* to its naval arsenal in the months prior to Pearl Harbor, the IJN had expended enormous resources building the super-battleships *Yamato* and *Musashi* to add to its fleet, reflecting the continued sway held by traditional "big gun" advocates in the imperial navy. In retrospect, Japan would have been better off building several more aircraft carriers in place of these eighteen-inch gunned giants. To be fair, however, such observations have the benefit of hindsight. Although all the world's leading navies recognized the tremendous progress made by naval aviation during the interwar period, the central role that the aircraft carrier would play in the Pacific was still far from clear prior to 1941. The IJN General Staff initially opposed the massed use of aircraft carriers in a preemptive strike on Pearl Harbor, which was essentially the brainchild of Isoroku Yamamoto, commander in chief of the Combined Fleet. The naval General Staff felt that such an operation was far too risky and preferred to adhere to traditional plans, which called for the Japanese battle fleet to lie in wait for the Americans as they advanced across the Pacific and then engage them in a classic sea battle closer to home waters. Thus, in a consensus-driven Japan, traditional views continued to exert influence at the institutional level, despite the rise of revolutionary new ideas in some quarters. The presence of the Japanese battleship fleet at the June 1942 Battle of Midway in the rear of Admiral Nagumo's carriers betrays a curious continuation of allegiance to big naval guns, despite the devastating impact of airpower on surface ships, as dramatically demonstrated by the IJN's own operations in the preceding six months.

Just as army traditionalists remained bound to the notion of infantry combat as their ultimate purpose in life, naval commanders found it difficult to discard the doctrine of fleet engagement as the central priority of IJN operations. Vice Admiral Gun-ichi Mikawa achieved a stunning victory over Allied cruisers in the night battle of Savo Island at the start of the Guadalcanal campaign but then withdrew without touching the transports, which still carried supplies to be off-loaded on

Guadalcanal. Japanese submarine doctrine called for their use against enemy warships rather than against merchant shipping. Although they achieved some significant victories over U.S. warships during the first year of the Pacific war, their use against commercial tonnage remained limited, despite promising results achieved along the east coast of Australia during 1942 and early 1943. By the same token, IJN airmen continued to favor warship targets over transports for much of the Guadalcanal campaign. In all these cases, an earlier appreciation of the strategic value of attacking enemy transport capacity would have done much greater damage to the Allied effort and might have seriously impeded the timetable of their counteroffensive in the South and Southwest Pacific, especially since the Allies suffered from a serious shortage of transport shipping during the first half of the Pacific war.

Japanese Capabilities versus the Allies'

The emphasis of Japanese army aviation on tactical support of ground forces and the IJN's priority of supporting battles at sea reveal an overly narrow focus on battle per se. Japan's recognition of the importance of logistical factors and other supporting elements in modern warfare proved woefully inadequate. Air transport and ferrying of aircraft from rear areas to the front, rapid airfield construction capabilities, development of effective radar, and installation of effective radio sets in fighter aircraft were all fields in which the Japanese were seriously deficient or lagged significantly behind their Allied opponents. The more primitive the field conditions, the more vital these factors were. Especially in New Guinea and the Solomon Islands, which became the main arena of combat with the United States and its allies from mid-1942 onward, the contest developed into one of logistics and support operations to sustain the combat effectiveness of the forces in the field. On several occasions during the Papuan campaign in eastern New Guinea in the autumn of 1942, navy *rikko* had to be diverted from combat operations to drop supplies to Japanese troops on the ground, since no transport aircraft were available for the task. This contrasts dramatically with the Allied feat of airlifting the U.S. 32nd Infantry Division from Australia to New Guinea by transport aircraft in the same period to face these same Japanese troops. In addition, for much of the Pacific war, both Japanese army and navy air units were periodically forced to take combat pilots out of the front lines so that they could fly replacement aircraft in from rear areas because of a lack of ferry crews.

The disparity in airfield construction capabilities between the Japanese and the Allies is well known. The Japanese, forced to rely largely on manual labor, typically required a month of backbreaking work to clear an airstrip out of the jungle; the Allies, using heavy mechanized equipment, could do so in a matter of days with a fraction of the manpower. Those airfields that the Japanese built or expanded from existing sites were often limited in the number of aircraft they could safely accommodate because of the Japanese lack of field engineering capabilities, which affected the extent of taxiways and revetments that could be constructed in a timely fashion. This restricted the Japanese ability to deploy large numbers of air units quickly to frontline areas, even when reinforcements were available, and it frequently resulted in dangerous congestion at forward airfields. The damage or destruction of more than 100 aircraft of the Japanese Fourth Air Army, parked wingtip to wingtip on the ground at Wewak and But on 17 August 1943 at the hands of the U.S. Army's Fifth Air Force, was a direct result of such congestion. This disaster was repeated at Hollandia, farther west along the northern New Guinea coast, on 30 and 31 March 1944.

On both those occasions, lack of adequate early warning was also to blame for the debacle. The Japanese had made some pioneering contributions to the development of radar in the prewar years, but they overlooked its significant military applications and failed to devote adequate resources to its development, thus falling seriously behind the West in radar technology. The Japanese did deploy ground-based early-warning sets to forward areas starting in the second half of 1942, and radar was in place at Wewak on 17 August 1943, but it consisted of a single set with a limited field of coverage. This inadequacy of the early-warning network was symptomatic of the Japanese war effort in the

Southwest Pacific. At Hollandia in March 1944, there was no radar network in place because the Japanese, faced with a shortage of available radar sets, were awaiting the arrival of the set from Wewak to provide the necessary coverage. This set never arrived at Hollandia, being lost en route when the ship carrying the equipment was sunk by American aircraft on 19 March 1944.

In the air, the lack of effective shortwave radios in Japanese fighter aircraft, allowing communication among pilots as well as with the ground, greatly hampered the coordination of formations in combat and often prevented Japanese fighters from taking full advantage of favorable tactical situations when they arose. For much of the Pacific war, aerial communication among Japanese fighters was limited to World War I–vintage visual methods, such as the rocking of wings or simple hand signals given from the cockpit. Ineffective radios, coupled with lagging radar development, also frustrated Japanese efforts at ground-controlled interception. It was not until the air defense of the Home Islands in 1944 and 1945 that the JAAF used such procedures extensively. In fact, the war ended just as the Japanese army was about to begin operating interceptor command centers of the sort the RAF already had in place by the Battle of Britain in 1940. The imperial navy never established a system of carrier-based control of airborne formations, as did its U.S. counterpart. Arguably, at the Battle of the Philippine Sea in June 1944 and at Leyte Gulf in October of that year, the IJN possessed the basic technology to do so. But at the time, its expertise in telecommunications had not matured to the point where it could deploy such a fully integrated system. By the time its abilities might have allowed such an operation, the imperial navy no longer possessed the carriers or the carrier-qualified aircrews to do so.

Japan's Priorities and Their Consequences

Japan's failure to pay due attention to the supporting elements of aerial warfare can be attributed both to the limited scope of the nation's total resources and to the lack of a unified organ of government responsible for national defense. With neither the material nor human abundance enjoyed by the United States and its allies, the Japanese army and navy were obliged to make some hard choices over the allocation of resources and the direction of development efforts. That they chose to focus so narrowly on effectiveness in battle at the expense of broader measures of success in modern warfare indicates a lack of foresight and limited strategic vision.

In the field of aircraft development, this is reflected in the absence of such protective measures as self-sealing fuel tanks and armor plate on the planes of both the JAAF and the IJN at the start of the Pacific war. The Japanese were well aware that they lagged several years behind the West in the development of more powerful aircraft engines. The United States was already designing aircraft with 2,000-horsepower engines by 1940, but it was not until 1942 that Japan had an operational engine of comparable power. Achievement of competitive performance with less power entailed the sacrifice of anything considered nonessential to lessen the weight of the plane. The Germans had learned some valuable lessons concerning aircraft protection in the Spanish Civil War, and most Luftwaffe bombers had self-sealing fuel tanks by the start of the Polish campaign in 1939. The Japanese, however, had drawn the wrong conclusions from their victories over the poorly armed Chinese in the late 1930s and underestimated the importance of aircraft protection in the face of strong opposition. It was not until losses rose sharply during the Guadalcanal campaign that the Japanese reconsidered. Before corrective measures could be taken, however, Japan's unprotected aircraft began to fall in increasing numbers. Japanese aircrews, which had never been numerous, fell with them, quickly thinning the ranks of experienced airmen down to dangerous levels.

Now their belated expansion of air training programs came back to haunt the Japanese. The United States had embarked on an ambitious expansion of air training for its armed forces by 1940, contemplating that if it were drawn into war, it would be a long-term struggle. Great Britain, already at war in Europe since 1939, was similarly engaged in turning out large numbers of airmen through its Empire Air Training Scheme and the British Commonwealth Air Training Plan. Japan, in contrast, realized that a long war would be to its disadvantage, because the superior manufacturing strength and manpower of its potential enemies would overwhelm Japanese forces over time. Therefore, Japan felt that it had no choice but to gamble on winning a short war, relying on a superior quality of forces. Although training programs expanded gradually in the years immediately prior to the start of the Pacific war, dramatic expansion of these programs did not take place until later. The IJN in particular continued to maintain highly selective training programs that did not begin to grow significantly until 1943. In the end, the initial margin of superior training and experience exhibited by its airmen proved insufficient to prevent serious attrition. The cadre of superbly trained airmen with which the IJN began the Pacific war was too small and lacked depth. As attrition mounted, particularly after the start of the Guadalcanal campaign in August 1942, the average quality of its airmen declined sharply; adequately trained reserves proved insufficient, and expanded training programs had yet to fill the gap. By the Battle of the Philippine Sea in June 1944, the superiority of the average Allied airman was absolute, as was the quality of the aircraft he flew and the centralized control procedures under which he operated. Faced with this formidable combination, the Japanese were confronted with the stark reality that, despite great sacrifice and huge losses, their air operations had become almost totally ineffective.

The JAAF was not quite as stringent as the IJN in its selection process, and it began to expand its training programs somewhat earlier than the navy did, but those programs still proved inadequate once the Pacific war began. As with the IJN over the Solomons in 1942 and 1943, attrition suffered over eastern New Guinea during the same period drained the JAAF of many of its most experienced pilots. The quality of its aerial opposition in subsequent campaigns over western New Guinea and the Philippines during 1944 was thus drastically reduced. In the end, for both the navy and army air forces, massed suicide sorties were the only option left to strike an effective blow in the face of overwhelming odds. Had the Japanese launched a more aggressive expansion of air training programs before the start of the Pacific war, even at the expense of a decline in average pilot quality, they would have been able to call on greater reserves of adequately trained airmen during 1942 and 1943, before Allied superiority in the air became insurmountable. It is unknown whether the availability of such reserves would have meant the difference between victory and defeat, but they certainly would have allowed Japan's air forces to sustain a higher level of combat effectiveness for a longer period during the crucial middle phase of the war in the Pacific.

As the products of the expanded wartime training programs began to reach frontline units, they found themselves at a disadvantage, not only in terms of experience but also in terms of equipment. Although self-sealing fuel tanks and armor plate began to appear on Japanese aircraft by 1943, the aircraft themselves were essentially the same machines that had equipped the JAAF and IJN at the start of the Pacific war.

Japan had undergone many years of Western tutelage in the design and manufacture of aircraft, beginning with a total reliance on the West for equipment in the years immediately before and after the First World War. There followed a transition phase in the late 1920s and early 1930s of licensed manufacture and modification of Western designs, often with the help of European engineers under contract. From the mid-1930s, the Japanese aircraft industry began to achieve self-sufficiency in the design and manufacture of its own airframes and engines. Then, in the latter half of the 1930s, Japan finally began to produce aircraft that were competitive with those from the West, but it was still heavily dependent on imported technology for components and subsystems. When the nation plunged into the Pacific war, it had produced only one or two generations of such aircraft. More significantly, after the successful military adoption of these aircraft in 1937–1940, Japan failed to initiate a significant number of successful follow-on designs. Had it done so, given an average development cycle of three years, these newer machines should have been ready for service during the crucial 1942–1943 period, when the outcome of the Pacific war hung in the balance. The next major wave of designs, however, did not come until 1940–1942. With the flow of Western technology largely cut off once the Pacific war began, some of these designs encountered serious technical problems, and their gestation periods were inordinately long. By the time these planes were ready for service use, the balance of power had shifted too far in favor of the Allies for them to have a significant impact on the war. Others failed to reach operational status before Japan's surrender.

The most notable example of this failure is the effort to develop a successor to the imperial navy's Zero fighter. Had the IJN issued specifications for a Zero replacement as that plane was coming into service in 1940, or even in the months prior to Pearl Harbor during 1941, it is conceivable that the fighter pilots of the imperial navy would have been flying the Zero's replacement in time to challenge the F4U Corsair in the Solomon Islands in 1943 or to face the F6F Hellcat in the Battle of the Philippine Sea in June 1944. But lulled into complacency by the Zero's outstanding success, and distracted by other demands on design personnel at the start of the Pacific war, serious attention to the development of a successor did not begin until April 1942. The Zero's chosen successor, the "Reppu" (A7M "Sam"), suffered numerous delays caused by technical aspects of the design and was still in the testing stage at war's end. The navy's reluctance to abandon the Reppu project also delayed the adoption of an alternative design in the form of the "Shiden" (N1K1-J "George"). As a result, the latter did not enter combat until the Philippines campaign in October 1944, and then only in small numbers. In combat, the Shiden suffered frequent problems with its engine and landing gear, making it less effective than had been hoped. A much improved version, the Shiden Model 21, popularly known as the Shiden-Kai, did not enter combat until March 1945, far too late to have any serious impact on the course of the war. Thus, the Zero, a design originally adopted for service in 1940, was never replaced as the IJN's primary fighter during the Pacific war. In contrast, the U.S. Navy's F6F Hellcat achieved a remarkable development cycle of less than two years. Initial design work on the F6F began in mid-1941; it entered squadron service in January 1943 and made its combat debut at the end of August that same year. Thereafter, it quickly replaced the F4F Wildcat as the U.S. Navy's main carrier fighter and demonstrated a decisive edge in performance over the imperial navy's Zero.

The U.S. aircraft industry achieved faster development cycles for major aircraft types and introduced them into large-scale service at a faster rate. Among major Japanese aircraft types whose designs were initiated in 1939 or 1940, only one, the JAAF's Type 3 fighter "Hien" (Ki 61 "Tony"), entered combat in any number by mid-1943. Others, such as the IJN's "Tenzan" torpedo bomber (B6N "Jill"), the "Raiden" interceptor fighter (J2M "Jack"), and the "Ginga" land-based bomber (P1Y "Frances"), did not enter operational service until the latter half of 1943 or the first half of 1944; nor did the "Suisei" dive-bomber (D4Y "Judy"), whose development had actually begun in 1938. In contrast, design work on many U.S. aircraft that went on to play major roles in the Allied victory were started in 1939 or 1940, and most entered squadron service during 1942 or 1943. These included the P-51, the P-47, the B-24, and the B-29. The design of the B-26, already in service at the time of Pearl Harbor, was also begun in 1939; the B-25, initiated in 1938, had reached squadrons even earlier during 1941.

Conclusion

There is certainly an element of truth to Vice Admiral Kuwabara's assertion that Japan began the Pacific war too soon. A miraculous postwar recovery and three decades of economic growth were needed before Japan could truly call itself a leading industrial power. But we have seen that technical and industrial inferiority alone cannot explain all the elements of Japan's aerial defeat in the Pacific war. Japan's military leaders knew from the start that they could not win a protracted war with the West. They reasoned that the one slim chance they had of reaching a negotiated settlement of the conflict was to maintain the initiative and keep the Allies off balance until they could consolidate their gains sufficiently to withstand the inevitable Allied counterattack. But Japan's defeat at the Battle of Midway and its misjudgment of the situation at Guadalcanal, which drew it into a fatal campaign of attrition, took place before the Allies had gained numerical superiority of forces. It is difficult to avoid the conclusion that the turning point in the Pacific war came not because of superior Allied numbers or material abundance but because of superior Allied operational skill. A narrow Japanese focus on battle, to the exclusion of other considerations, as well as doctrinal inflexibility, played a major role in this early reversal of Japan's fortunes. Once the initiative had passed into Allied hands, the lack of cooperation between the imperial army and navy and the Japanese failure to fully mobilize national resources for an all-out war prior to 1943 simply aggravated the numerical and technical superiority of their Western foes.

Suggestions for Further Research

The literature on the Japanese aerial effort in World War II is rich and varied, yet there are important questions and avenues of research that have been neglected. Despite several general economic and military histories of the Japanese war effort, the definitive examination of Japan's wartime aviation industry is yet to be written. Research into performance specifications and design solutions would tell us much about Japanese strategic, engineering, and managerial thinking. The far more complex topics of general Japanese procurement planning and production management also remain to be explored. These issues are as much about Japanese society as about the armed forces of imperial Japan.

In a more narrowly military sphere, much remains to be discovered about Japanese views on the support of aerial forces and air operations. Training of both air and ground crews, the development of technology and engineering, airfield construction and basing methods, and the assessment of wastage and corruption at all levels are just a few of the topics beckoning researchers who are interested in a broader understanding of how the Japanese waged war in the mid-twentieth century.

These questions are related to larger issues of national culture and historical traditions. Clearly, the mind-set of the samurai persisted even after the official dissolution of that social class in the nineteenth century, as evidenced by the chronic aversion to the more mundane and material aspects of war making in the industrial age, such as logistics, protection of lines of communication, maximization of transport infrastructure, mobilization of the population and industry, and other "inglorious" duties. The case can be made that the Japanese, facing daunting odds because of the vastly superior scale of the populations and economies of their opponents, needed to maximize their productive and managerial efficiency but often did the opposite, thus widening the chasm between them and their opponents. What were the causes of such apparently counterproductive resource management?

When an army not only struggles to supply its forces in the field but also fails to give proper attention to the methods and systems of supply, it may indicate something beyond poor decisions by the military leadership. What was the relationship between the military and general society in imperial Japan that led to such consequences? Why didn't the Japanese armed forces better mobilize the scientific and technological expertise available in Japanese academia, which included some of the world's leading scholars and theorists? Is there a connection between the war's outcome and Japan's amazing economic and scientific turnaround in the postwar years—the much marveled at "Japanese miracle"? These are questions worthy of further thought, research, and analysis.

Recommended Reading

The Pacific war was one of unparalleled scope, cultural antithesis, and technological marvels, and one result has been a remarkable breadth and variety of historical literature. The serious student of Japanese airpower and its defeat in World War II can refer to the official histories of the major belligerents, some illuminating personal memoirs and retrospectives by participants, and more traditional histories of operations, airpower theory and planning, and aircraft design and production.

A good start for understanding the overall air war in the Pacific from the Japanese perspective is Masatake Okumiya and Jiro Horikoshi, *Zero!* (New York: Ballantine Books, 1956). Though the authors were architects of the Japanese navy's air arm, they write expansively about general Japanese views of airpower, including their analysis of what went wrong. This should be supplemented with a study of Mark R. Peattie,

Sunburst: The Rise of Japanese Naval Air Power, 1909–1941 (Annapolis, Md.: Naval Institute Press, 2001), a comprehensive look at the development of the imperial navy's air arm. Also of interest in understanding the rise of Japanese naval airpower is the examination of the Royal Navy's influence in Arthur Marder, *Old Friends, New Enemies: The Royal Navy and the Imperial Japanese Navy,* 2 vols. (New York: Oxford University Press, 1981–1990). Perhaps the best account of Japanese naval aviation from the inside can be found in the memoirs of Japanese fighter ace Saburo Sakai, with Martin Caidin and Fred Saito, *Samurai!* (New York: E. P. Dutton, 1956). See also my own *Imperial Japanese Naval Aviator, 1937–45* (Oxford: Osprey Publishing Limited, 2003), no. 55 in the Osprey Warrior Series.

Japan's rise to world power status in military aviation by the time of Pearl Harbor must be seen in the context of general military developments in imperial Japan. Excellent works that provide background on the development of the Japanese navy in this period are David C. Evans and Mark R. Peattie, *Kaigun: Strategy, Tactics, and Technology in the Imperial Japanese Navy, 1887–1941* (Annapolis, Md.: Naval Institute Press, 1997), and H. P. Wilmott, *The Barrier and the Javelin: Japanese and Allied Pacific Strategies, February to June 1942* (Annapolis, Md.: Naval Institute Press, 1982). Americans' perception of Japanese strategic thinking and their response to it are fully outlined in Edward S. Miller, *War Plan Orange: The U.S. Strategy to Defeat Japan, 1897–1945* (Annapolis, Md.: Naval Institute Press, 1991). On the Imperial Japanese Army and its aerial forces, see Saburo Hayashi, with Alvin Coox, *Kogun: The Japanese Army in the Pacific War* (Quantico, Va.: Marine Corps Association, 1959), and Coox's in-depth examination in *Nomonhan: Japan against Russia, 1939,* 2 vols. (Stanford, Calif.: Stanford University Press, 1985). Still valuable are the essays by Asada Sadao (on the Japanese navy) and Akira Fujiwara (on the Japanese army) in Dorothy Borg and Shumpei Okamoto, eds., *Pearl Harbor as History: Japanese–American Relations, 1931–1941* (New York: Columbia University Press, 1973).

The most comprehensive and analytical history of the war in the skies over the Pacific is found in Richard J. Overy, *The Air War, 1939–1945* (New York: Stein and Day, 1980), but the history of aerial operations is also covered in the official U.S. service histories. The role of the U.S. Army Air Forces in the defeat of Japanese airpower is examined in Wesley Frank Craven and James Lea Cates, eds., *The Army Air Forces in World War II,* 7 vols. (Chicago: University of Chicago Press, 1948–1953). Samuel Eliot Morison's majestic fifteen-volume *History of United States Naval Operations in World War II* (Boston: Little, Brown, 1947–1952) has much relevant material in the eleven volumes devoted to the war in the Pacific. The best single-volume history of the Pacific campaigns remains Ronald Spector, *Eagle against the Sun: The American War with Japan* (New York: Macmillan, 1985). Other general histories of World War II contain some insights into the Japanese–American air war. The best in recent years are Richard Overy, *Why the Allies Won* (New York: W. W. Norton, 1995), and Williamson Murray and Allan R. Millett, *A War to Be Won: Fighting the Second World War* (Cambridge, Mass.: Belknap Press of Harvard University Press, 2000). Also, for a discussion of how the defeat of Japanese naval aviation factored into the war's outcome, see Masanori Ito, with Roger Pineau, *The End of the Imperial Japanese Navy* (New York: W. W. Norton, 1956).

The histories and personal accounts of individual battles and campaigns are too numerous to list, and many of them shed light on some aspects of Japan's aerial defeat. Also copious are accounts of the strategic bombing campaign against Japan, but these frequently concentrate on American plans and operations and on the general consequences for the Japanese without directly addressing the defeat of Japanese airpower.

Japanese aircraft and the aviation industry are also covered in the literature. A standard reference work is René J. Françillon, *Japanese Aircraft of the Pacific War* (Annapolis, Md.: Naval Institute Press, 1987). Also helpful is Robert C. Mikesh and Shorzoe [*sic*] Abe, *Japanese Aircraft, 1910–1941* (Annapolis, Md.: Naval Institute Press, 1990). There are a number of books on specific aircraft, such as the Osprey Combat Aircraft Series, including my own no. 22, *Mitsubishi Type 1 Rikko "Betty" Units of World War 2* (Oxford: Osprey Publishing Limited, 2001). Also noteworthy is the published wartime intelligence bulletin of the U.S. Army Air Forces: Air Force Historical Foundation, *Impact: The Army Air Forces' "Confidential" Picture History of World War II,* 8 vols. (New York: James Parton and Co., 1980). One can also consult the U.S. Strategic Bombing Survey report, *The Japanese Aircraft Industry* (Washington, D.C.: GPO, 1947), for further details on aircraft production. For some of the wider implications of the development of the Japanese aircraft industry, refer to Richard Samuels, *Rich Nation, Strong Army: National Security and the Technological Transformation of Japan* (Ithaca, N.Y.: Cornell University Press, 1994).

For those who can read Japanese, there are a multitude of additional sources. Of primary interest is Nihon Kaigun Kokushi Hensan Iinkai, ed., *Nihon Kaigun Kokushi* [History of Japanese Naval Aviation], 4 vols. (Tokyo: Jiji Tsushinsha, 1969). Minoru Genda, a Combined Fleet air staff officer and the planner of many important air operations, has written a two-volume history of the naval air war, *Kaigun Kokutai Shimatsuki* [A Record of the Japanese Naval Air Service] (Tokyo: Bungei Shunju, 1961–1962), as well as his own account of the Pearl Harbor raid, *Shinjuwan Sakusen Kaikoroku* [Recollections of the Pearl Harbor Operation] (Tokyo: Yomiuri Shimbunsha, 1972). Another naval aviation memoir is Kuwabara Torao, *Kaigun Koku Kaisoroku: Soso Hen* [Recollections of Naval Aviation: Early Years] (Tokyo: Koku Shimbunsha, 1964). Iwaya Fumio has written a two-volume history of land-based naval bombers, *Chuko: Kaigun Rikujo Kogekikitai Shi* [The Medium Bomber: A History of the Navy's Land-based Attack Aircraft Force] (Tokyo: Shuppan Kyodosha, 1956–1958). Imperial army aviation has garnered less attention, but there is one account of its defeat: Koku Hiho Sankai, ed., *Rikugun Koku no Chinkon* [Requiem for Army Aviation] (Tokyo: Koku Hiho Sankai, 1978).

Of great use are the Japanese official histories, the *Senshi Sosho* [War History] series of more than 100 volumes undertaken by the Boeicho Boeikenshujo Senshibu (Defense Agency, Defense Training Institute, War History Department; formerly known as the "War History Room" of the institute). Many of the volumes are relevant to a study of Japanese military aviation in World War II, but several are dedicated specifically to the subject. These include the three-volume series *Rikugun Koku no Gunbi to Unyo* [Army Aviation Armaments and Their Use] (1971–1976); *Rikugun Koku Heiki no Kaihatsu, Seisan, Hokyu* [Development, Production, and Supply of Army Aviation Armaments] (1975); *Rikugun Koku Sakusen Kiban no Kensetsu Unyo* [Construction and Employment of the Operational Foundations of Army Aviation] (1979); *Kaigun Koku Gaishi* [Historical Overview of Japanese Naval Aviation] (1976); and three volumes on aerial campaigns—Home Islands defense (1968), eastern New Guinea (1969), and Manchuria (1972). Many of the operational and campaign histories have substantial sections on aerial strategy and operations.

CHAPTER SEVEN

Defeat of the Luftwaffe, 1935–1945

James S. Corum

By almost any measure, the Luftwaffe was superior to its enemies in 1939–1941. The Me 109 fighter was superior to most opponents, and only Britain's Spitfire could match it in combat. The Ju 88, He 111, and Do 17 medium bombers were some of the best machines of their day. In Poland, Norway, France, North Africa, and Russia, the Ju 87 Stukas proved to be fearsomely effective as close support aircraft. In addition to combat machines, the Luftwaffe could field 500 transport planes, the largest air transport force in the world, and one that played a decisive role in several early campaigns. The Luftwaffe also had a superbly trained paratroop–glider division that brought a new intensity to warfare by its ability to seize key positions hundreds of miles behind the front lines.

The Luftwaffe of 1939–1940 was a well-rounded force, able to defend the homeland, win air superiority, support advancing forces, and conduct strategic bombing campaigns. Contrary to popular myth, the Luftwaffe put considerable effort into creating a strategic bomber arm and doctrine. It was the first air force to employ "pathfinder" bombers and was the only one able to attack enemy targets accurately at night in 1940—as it proved in the massive nighttime attack on Coventry in November of that year. Although it was the newest air force among the major powers, it also had the most combat experience, thanks to the war in Spain, where 20,000 Luftwaffe personnel served between 1936 and 1939. Generally, Luftwaffe pilots were better trained than their opponents.

Germany also had the advantage of a first-class aircraft industry with excellent designers and engineers and a huge, modern industrial base (the second largest economy in the world in 1939) with a highly skilled workforce. Between 1939 and 1945, German aeronautical engineers would astound the world with the first jet fighters and bombers, rocket planes, and cruise and long-range missiles.

Remarkably, none of the German advantages lasted for long. By 1941, the Luftwaffe was heavily outnumbered on every front. By 1942, few of the Luftwaffe's aircraft had an edge on their opponents. The Allies soon outstripped the Luftwaffe in both aircraft numbers and quality. The well-trained personnel of 1939–1941 were lost to attrition and replaced with aircrews that were inferior to their British and American counterparts. By 1942–1943, the Allied powers had become adept at using air transport and close air support—even as those skills declined among the Germans. By the midpoint of the war, the Luftwaffe was locked into a steady and irreversible decline. This chapter explains how the Luftwaffe, through bad leadership and misguided strategy, threw away its initial advantages and lost the war. Of course, Allied numbers and industrial capability played a central role in defeating the Luftwaffe, and one could argue that the Germans ultimately would not have prevailed against such a mighty coalition as the United States, the Soviet Union, and the British Empire, even if they had made no grand mistakes. However, it is fair to point out that the Germans played a decisive role in their own defeat.

Doctrinal Failure

Germany's first great defeat in World War II, the failure to overcome Britain in the summer of 1940, was largely rooted in a deficient air-war doctrine.

The popular view of the Battle of Britain is that it was a close contest. It is argued that if the Luftwaffe had kept up its campaign against the Royal Air Force (RAF) sector airfields, it could have gained air superiority over southern England, which would have made a landing by the army possible. The reality was different. In 1940 the Luftwaffe was fighting against Britain's two greatest strengths—the best air defense system in the world and the powerful Royal Navy. The Luftwaffe's small advantage in single-engine fighters was not nearly enough to overcome the RAF—unless the RAF made some enormous mistakes. Even if the Luftwaffe had won, it is unlikely that the small German navy could have successfully landed the army on the English coast in the face of the Royal Navy. Moreover, the Luftwaffe had only a few weeks to win the air superiority battle before bad weather made an invasion impossible. The Allied air campaign over Germany proved just how difficult is was to win air superiority against a determined foe—even when the Allies possessed a greater margin in aircraft numbers and trained aircrews than the Luftwaffe had in 1940.

Britain's great vulnerability in 1940 was its dependence on seaborne transport to feed its people and supply raw materials to industry. If the sea lifeline had been cut, Britain would have had no alternative but to make peace with Germany. Yet the Luftwaffe lacked a naval strike force that was trained and equipped to attack British shipping and ready to work with the U-boats to exploit Britain's vulnerability. The absence of an effective naval air arm was not due to any shortage of aircraft or technology—it was due solely to a lack of doctrine rooted in interservice rivalry.

In World War I, the German navy had possessed an effective air arm, which in 1917–1918 won air superiority over the English Channel and successfully attacked and sank some Allied ships. When Germany rearmed, the navy wanted to rebuild an air arm capable of fleet reconnaissance and assaults on enemy ships. However, Hermann Göring, as commander in chief of the Luftwaffe, insisted that Germany could have only one air force. Everything that flew must come under the command of "his" Luftwaffe. Göring promised the navy an air arm that would support the fleet and strike enemy shipping and serve under the navy's operational control. But Göring's promise, like so many others, proved empty.

Before the war, the navy lobbied for a force of long-range bombers, able to fly far out to sea and strike shipping with bombs and torpedoes. Such a force was easily attainable with Germany's resources. The He 111 and Ju 88 bombers were well suited for such tasks, and senior Luftwaffe commanders, such as General Helmut Felmy, commander of Luftflotte II, argued that Germany needed a strong naval air arm for antishipping strikes and to mine ports in case of war with Britain. Göring ignored the navy and his own senior commanders and put the creation of an air force for a rival service at the bottom of his priority list. In 1939 Germany had only a weak naval air force of fewer than 250 aircraft, mostly obsolete seaplanes and flying boats for reconnaissance. In the fall of 1939 Admiral Raeder pleaded with Göring for a few wings of Ju 88 bombers to be equipped and trained for naval operations. He was ignored, however, and naval air operations remained a low priority.

Before 1939, Germany developed a new weapon that might have had a decisive impact on Britain's sea trade—the air-dropped magnetic mine. Unlike the naval contact mines of World War I, the magnetic mine was hard to detect and sweep. Sown at night in major shipping channels, it would detonate when a large mass of metal, such as a ship, passed nearby. If Germany had waited to accumulate a large store of these mines and sown them in the approaches of all the major British ports, British trade could have been crippled. Instead, in 1939 and early 1940, Göring refused to allocate bombers to mine laying, and the small naval air arm was able to drop only a few mines in small raids. Used in such a fashion, the new weapon sank a few ships and created a stir, until one was recovered and examined and effective countermeasures were developed.

Lacking long-range antishipping bombers at the start of the war, the Luftwaffe improvised such a force from a handful of four-engine FW Condor passenger planes. Modified FW 200s had great range, which meant that they could fly far out to sea, outside the range of home-based RAF fighters. However, the planes were not really suitable as strike aircraft; their light structure was not easily modified to carry an adequate load of armor, defensive guns, and torpedoes. Nevertheless, by August 1940, a force of thirty FW 200s was ready to fly antishipping strikes into the Atlantic under naval operational control. The results from this force, which never had more than a dozen planes available, were impressive. Between August 1940 and February 1941, the FW 200s sank eighty-five Allied merchant ships, for a total of 363,000 tons. It was an astounding performance. Yet the Luftwaffe was still reluctant to learn from this success. Its maritime strike capability languished, and only in 1942 did the Luftwaffe create a few squadrons of He 111 torpedo bombers, which caused significant damage to Allied convoys in the Arctic Sea and the Mediterranean.

Luckily for the Allies, the Luftwaffe never invested much effort or interest in the naval air war. One can only imagine the damage the Luftwaffe might have done to Britain if a force of 100 to 200 long-range strike aircraft had been available at the start of the war or if the Luftwaffe had combined air attack with a large-scale mining campaign.

Failures in Logistics and Support Planning

When the war began in 1939, the Luftwaffe had a highly developed mobile logistics system geared toward supporting short blitzkrieg campaigns. Although German bombers had the range to strike Poland, France, or Britain from German bases, most of the Luftwaffe's fighters, Stukas, and reconnaissance planes had an effective range of only 100 to 200 miles. If the Luftwaffe were to support mobile army operations, it needed the capability to jump forward quickly, put captured airfields into operation, and keep its planes supplied with bombs and fuel. As armies advanced, short-ranged Luftwaffe aircraft could cover the front units, assure air superiority, and provide effective close air support. Moreover, operating from airfields near the front lines maximized sortie rates. In the battle for France in 1940, Luftwaffe fighter and Stuka units averaged an impressive four to six sorties a day, whereas the French air force fighters averaged only one per day.

In September 1939 the Luftwaffe possessed 117 airfield and engineering companies and motorized supply columns to support the air fleets and corps. As the army advanced, the airfield companies would move into a captured airfield, effect repairs, and open the field for operations within twenty-four hours. Motorized supply units advanced behind the armored spearheads, bringing fuel and munitions forward. The system was successful in supporting the Luftwaffe's forward units, although even this relatively large support force was strained by the campaign and relied heavily on transport planes to keep supplies flowing to the airfields.

After the Polish campaign, the Luftwaffe expanded its logistics organization. The supply and transport system played a major role in the Luftwaffe's superb performance in Norway, the Low Countries, and France in 1940. The high sortie rate of Luftwaffe fighter and Stuka units gave the Germans a decisive advantage. By flying more sorties per day than the French or British, the slight numerical superiority in aircraft enjoyed by the Luftwaffe in 1940 was translated into a vastly superior force over the battlefield.

The major flaw in this mobile logistics system was that it was designed only for short campaigns. Although the Luftwaffe's airfield units could keep forward units supplied with fuel and bombs, the groups and wings had only a minimal capability to repair and rebuild aircraft. If an aircraft needed major repairs, it had to be loaded onto a truck or railcar and shipped back to Germany, where it would be repaired or rebuilt at the factory. This lean repair and maintenance infrastructure saved the Luftwaffe money, but it also meant that damaged aircraft were out of action for a long time. The system worked in the short campaigns of 1939 and the spring of 1940, when the Luftwaffe could throw every available aircraft into the battle, win quickly, and rebuild the force afterward. However, if the campaign ever became a war of attrition, the lack of forward maintenance and repair units guaranteed that aircraft serviceability rates would drop precipitously. Starting with the Russian campaign in 1941, this is precisely what happened.

The Wehrmacht's planning for the campaign in Russia is a prime example of strategic hubris. The German high command operated under the assumption that the massive Soviet armed forces would simply collapse, and victory would come in a matter of weeks. The Wehrmacht assumed that the supply arrangements already in place were adequate for a short campaign, so there was no planning for the possibility that the Soviets could hold on into the winter and force the army and the Luftwaffe to fight more than a thousand miles from their home bases, backed by a tenuous supply line. Professional soldiers should have known better. Indeed, even before the start of the Russian campaign, the Luftwaffe's logistics and repair system proved barely capable of handling the campaigns in the Balkans and North Africa.

In the Balkans campaign in April 1941, the VIII Air Corps started out with serious logistics problems. The air corps was based in Bulgaria, and the poor road and rail network made it difficult to supply fuel, bombs, and parts for several hundred aircraft. When the Wehrmacht invaded Greece, the VIII Air Corps, per standard Luftwaffe doctrine, moved fighter and Stuka units forward 100 miles onto a captured Greek airfield. However, Greek roads were so poor that the mobile supply columns were unable to bring up enough fuel and munitions. So the VIII Air Corps kept its forward task force flying only by the fortuitous capture of British fuel and by using every available transport plane to bring supplies forward. The Luftwaffe was able to provide effective support to the army in the Balkans—but just barely. The problems of early 1941 foreshadowed the task the Luftwaffe would face in Russia. There, it was not a matter of jumping units 100 miles forward; the Luftwaffe would have to jump 1,000 miles forward—and do it over poor roads and a different-gauge railroad track system.

The Wehrmacht's strategy in the 1941 Russian campaign has been debated for decades. One question is whether the Germans could have taken Moscow if more forces had been allocated to the central army group. A more pertinent question is, how did the Germans manage to get so far with their shoddy logistics planning and infrastructure? By the time the Wehrmacht reached Moscow in October 1941, it had only a handful of working tanks and trucks and hardly enough fuel to move them. The greater part of the army's equipment had broken down from the rigors of the campaign or was out of fuel. It was the same for the Luftwaffe. By late 1941, less than 30 percent of the Luftwaffe's forward air units were operational. Airfields were desperately short of fuel, and each of the air fleets had hundreds of aircraft awaiting repair. The air support that the army had come to expect in short campaigns over relatively short distances was not available at the vital moment when the Russians successfully counterattacked in December 1941.

Throughout the war, and on every front, Luftwaffe units generally had an aircraft serviceability rate of 50 to 60 percent. In contrast, the Allies created a large maintenance and repair system for forward air units. Because of better maintenance facilities, the RAF and U.S. Army Air Force normally had 70 to 80 percent of their assigned aircraft operational. It was common for heavily damaged British and American planes to limp home and crash at their home bases—and to see the same planes fixed and flying again in a few weeks. The Allies' superior repair and maintenance system became a key advantage in the battles of 1942–1944.

As the tide of war turned against Germany, the Luftwaffe's weak repair system ensured the loss of thousands of aircraft. As the Allies advanced into Tunisia, Sicily, and southern Italy, they found hundreds of only slightly damaged Luftwaffe aircraft that had been abandoned. Many planes needed only minor repairs to make them flyable, but the Luftwaffe's inability to do this near the front ensured that these aircraft became a total loss. During the Russian summer offensive of 1944 and the Allied advance across France, the story was the same. The Allied advance on two fronts forced the Luftwaffe into a headlong retreat, with no time to salvage planes because of its limited maintenance resources. Some of the overrun Luftwaffe airfields resembled vast junkyards. Indeed, as much as a third of the Luftwaffe's aircraft losses in 1943 and 1944 were due to the planes being abandoned.

Mismanagement of Aircraft Production

The failure of the German war economy has been discussed in detail in several works, so only a few key points will be highlighted here. First of all, Germany had enormous capability to produce aircraft. When rearmament began in earnest in the 1930s, Germany had a highly sophisticated economy and a world-class aircraft industry. Aircraft firms such as Junkers, Dornier, Heinkel, Focke-Wulf, and Henschel had a cadre of experienced designers and engineers. Germany had a large, well-educated technical workforce. As shown by the many revolutionary aircraft designs produced during the war—including the world's first jet fighter, rocket planes, and cruise missiles—Germany was significantly ahead of the competition in many areas of aeronautics. The German failure lay in the inability to translate good ideas and prototypes into mass production.

During the first phase of rearmament from 1933 to 1936, General Wilhelm Wimmer led the Luftwaffe's Technical Office, which was responsible for developing and procuring aircraft. The first generation of aircraft designed and procured consisted of such unimpressive designs as the Heinkel 51 and Arado 68 fighters and the Dornier 23 bombers. However, these aircraft had already been developed and could be built quickly to equip new air force units until better aircraft became available. Wimmer oversaw the development of the second generation of Luftwaffe aircraft—the He 111 and Do 17 bombers, the Me 109 fighter, and the Ju 87 dive-bomber, all of which went from design competition to prototype to production within three years. These were among the most advanced aircraft of the era and were equal or superior to anything produced by Germany's rivals. Wimmer, considered by his colleagues to have the best technical mind in the Luftwaffe, laid the foundation for the Germans' early success in the air war. In 1936, however, Göring removed Wimmer and replaced him with his old friend and Great War flying ace Ernst Udet—an action that would be decisive in the Luftwaffe's ultimate failure.

Udet was completely unsuited to be the chief of design and production. He was a highly skilled pilot who had made his living as a stunt pilot in the 1920s and was Germany's greatest living ace (sixty-eight kills in the war). He probably would have been happiest as the Luftwaffe's chief test pilot, but he had no experience in management or industrial production and little interest in the large amount of paperwork and planning that went with the job. In fact, one of his decisions crippled German aircraft development. Udet was a great believer in dive-bombing as the only accurate means of putting bombs onto a target. He had tested dive-bombers and was enthusiastic about their potential, and he was supported in this view by the Luftwaffe's chief of operations, Hans Jeschonnek (who became Luftwaffe chief of staff in 1939). Udet directed design teams to make *all* bombers capable of dive-bombing, including two- and even four-engine machines. This pronouncement meant that work on highly promising advanced bomber designs, such as the Ju 88 fast bomber, was brought to a halt so that the design teams could modify the aircraft for the additional stresses of dive-bombing. The Ju 88 medium bomber program was delayed for more than a year, as was the Dornier Do 217, intended to replace the Do 17. The He 177 bomber, under development before the start of the war, was hampered by the requirement to adapt a heavy, high-altitude bomber for low-level dive-bombing.

Under Udet, the third generation of Luftwaffe aircraft was either delayed in development and production or of poor design. Many, if not most, aircraft projects never get past the initial design and prototype stages. One of the most important requirements for the chief of design and production was to weed out bad designs before they made it to full-scale production. Udet was generally unwilling to kill off bad projects even after it became clear that the designs were hopelessly flawed. For example, the He 177 heavy bomber was designed to operate at high altitude and carry a large bomb load. In most respects, it was theoretically superior to the American B-17 or the British Lancaster. The problem with the He 177 was the engine layout. To cut aircraft drag, the engines were placed in tandem, and two engines in each wing nacelle powered one propeller. This resulted in a very clean aircraft design, but there was one major drawback: the two large engines linked together could not be cooled effectively, and the resulting heat buildup caused numerous engine fires. This problem was never solved and should have caused the program to be abandoned. In 1942, however, after years spent trying to make a poor design work, the Luftwaffe was desperate for a heavy bomber, so serial production was ordered, and more than 1,000 of the planes were produced. Despite this massive effort, few He 177s ever saw combat, mainly because the engines broke down or caught fire.

A senior officer can be forgiven for one bad design project, but Udet had several. The Me 210, designed to replace the excellent Me 110 heavy fighter, was a mechanical nightmare, and the production line shut down soon after starting. The Henschel (Hs) 129, designed as a ground attack plane and the successor to the Ju 87 Stuka, saw its first action in 1940. Again, the design concept was quite advanced. The pilot and engines were encased in heavy armor, which made the plane virtually impervious to ground fire. In addition, the Hs 129 could carry heavy cannon for tank busting. Yet the first prototypes proved to be a failure in combat. The plane was so underpowered that it was almost impossible to fly with a full armament load, and poor cockpit layout hindered the pilot. After a complete redesign, it finally entered production and combat in 1943, where it earned a reputation as a deadly but mechanically cranky aircraft with a low serviceability rate. Indeed, the only truly successful aircraft project under Udet's tenure was the FW 190 fighter, which is acknowledged as one of the best fighter-bombers of the war.

The German aircraft and engine industry was poorly structured to fight a long, total war. Before the war, even the newest German aircraft factories were small compared with the British and American ones. Although a large number of small factories made the industry less vulnerable to grand strategic bombing, it also prevented the Luftwaffe from employing the most efficient methods of mass production. Throughout the war, the per-worker productivity in German aircraft factories lagged significantly behind that of British and American workers. Another problem was simply too many programs with too little guidance. Aircraft and weapons were developed and produced in small batches, while early in the war, a large part of the German engine and heavy industry plant lay virtually idle.

Although Germany began the war with the most effective combat air force of the major powers, the initial advantage in numbers and quality quickly slipped away. In 1939 Germany produced an impressive 8,295 aircraft, but in 1940 this increased to only 10,247, and in 1941 the number was only 11,247. As early as 1940, Britain was producing more aircraft than Germany. Yet in 1940–1941, neither Göring nor Udet nor Jeschonnek saw any urgent need to reform the aircraft industry and dramatically increase production. The possibility of masses of modern American aircraft flowing to the Allies or even of the United States entering the war did not disturb the German leadership. Although the huge industrial potential of the United States was well known, Hitler, Göring, and the Nazi inner circle considered it irrelevant. They believed that Germany was sure to beat both Britain and the Soviet Union by 1941, and after that, the United States would have to face Germany alone, with the latter possessing the entire industrial potential of Europe.

Finally, the moribund state of the German aircraft industry and Udet's mismanagement of it could not be ignored, so Göring placed aircraft production under the far more capable direction of Field Marshal Ehrhard Milch of Lufthansa. Unlike Udet, Milch had considerable experience in the aircraft industry and was a highly competent manager. He instituted an array of reforms and efficiency measures that boosted German aircraft production to over 15,000 in 1942 and 25,000 in 1943. Such measures, however, were undertaken far too late to meet the enormous demands of a war of attrition. German aircraft production peaked in 1944 with an impressive total of 40,000, but the same year, the Americans alone produced 96,000 airplanes. With the additional Russian, British, and British Commonwealth aircraft production at the height of the war, the Allies were outproducing Germany by a factor of four to one.

Richard Overy, in *Why the Allies Won,* notes that the Germans started with many advantages, including "a wealth of resources, a large class of competent entrepreneurs and engineers and a highly skilled workforce." Yet the German economy, on the whole, performed very poorly. As Overy notes, "it was not enough of a command economy to do what the Soviet system could do; yet it was not capitalist to rely, as America did, on the recruitment of private enterprise" (205).

Failure to Support Coalition Air Forces

The Luftwaffe did not fight the war alone. From 1940 to 1944, German forces were in action beside those of Italy, Romania, Hungary, and Finland. In addition, Bulgaria, Croatia, and Slovakia all provided forces to support German operations. Japan was technically an ally, but the vast distance separating the two nations and the lack of common interests meant that there was little economic or military coordination between them. The most important of Germany's allies (Italy, Romania, Hungary, and Finland) all had considerable resources, industrial capacity, and military capabilities that could have contributed significantly to the air war. Yet Germany, as the coalition leader, consistently pursued policies that ensured that its allies remained militarily and technologically weak—even when Germany desperately needed help on every front.

Germany led a coalition of reluctant allies, drawn into the orbit of the Third Reich in the 1930s when small, central European nations had to side with either the Soviet Union or Germany. They chose Germany as the lesser of two evils. Only Italy had any real ideological affinity with Germany. Yet Italy entered the war late, in June 1940, and then only in the hope of a quick victory. Mussolini had overridden the advice of his senior officers, who argued that Italy was in no economic or military position to fight a war. However, Mussolini was able to bring a large air force and navy into the war at Germany's side.

Three of Germany's allies—Italy, Romania, and Hungary—had aviation industries with the capability to design and build aircraft. The Romanian and Hungarian industries were fairly small and limited to building mostly foreign-designed engines and aircraft under license, but Italy possessed one of the largest aircraft industries in the world in 1940.

Before the war, the German policy was to actively discourage potential allies from building up modern defense and aircraft industries. Göring, director of Germany's economic program as well as commander of the Luftwaffe, envisioned a future in which the small nations would serve as suppliers of food and raw materials to Germany, whose industry would supply the manufactured goods.

During the 1930s, Hungary and Romania sought to build modern air forces using licensed modern German aircraft and engines, but they were denied by Göring. In 1941, when the Romanians went to war alongside the Germans, their IAR 80 fighters were effective aircraft for the Russian front.

Italy had one of the largest air forces and industries in the world in 1939 and, on paper, looked like a formidable airpower. In fact, military adventures in Ethiopia and Spain had bankrupted Italy. Most Italian air force planes were obsolescent, and the aircraft industry was grossly undercapitalized and unable to mass-produce aircraft. The greatest weakness of the Italian aircraft industry was in engine development and manufacture, which lagged far behind that of the British, Americans, and Germans. Italian aircraft companies such as Fiat and Macchi had excellent engineering teams capable of designing first-rate airframes, but the problem was to produce engines with enough horsepower to turn good airframes into first-class fighting machines. Try as it might, the Italian engine industry could not design and produce the high-performance engines required for modern warfare. When Germany and Italy became formal allies in 1938, the Luftwaffe expressed little interest in strengthening or improving the Italian air force, even though the Germans were well aware of the many weaknesses of their ally.

Parts and repair facilities were in short supply, so from the start of the war, Italian air force units suffered from an extremely low operational rate of only 30 to 40 percent of assigned aircraft. Only a few months into the war, it became clear that Italy would require considerable help from Germany to wage war effectively in the Mediterranean. Little help would come.

In late 1940 the Italian army suffered a catastrophic defeat in North Africa at the hands of the British. Although the Italian air force outnumbered the British by a two-to-one margin, the British won air superiority with their better Hurricane fighters. To stave off total Italian collapse, the Germans rushed a few army divisions and the X Air Corps to North Africa in early 1941. Although the Germans and Italians maintained liaison officers at the higher headquarters, cooperation between the two air forces remained fairly ineffective. At the lower levels of command, they worked together informally, with Italian fighters flying escort for German bombers, and German Stukas providing close air support for the Italians. In contrast to the low regard the Germans had for the Italian army, they praised the bravery and dedication of the Italian airmen who flew beside them in North Africa. The Italians did their best, but with inferior aircraft and logistics support, their best was not very good.

If the Germans had been able to provide technical and financial assistance, the Italians could have had an effective air force. But the Germans provided little help to Italian manufacturers, and by 1942, Alfa Romeo was able to deliver only fifty engines a month. When the Italians tried to buy German aircraft engines, the Luftwaffe made only a handful available.

When Germany invaded Russia on 22 June 1941, it had less than 3,000 aircraft available, fewer than it had to support the invasion of France in May 1940. In Russia it was a case of too much front and too few aircraft. Although the Luftwaffe performed brilliantly in the early stages of the campaign, it simply lacked sufficient numbers of aircraft to carry out its missions. This meant that the Hungarian, Romanian, and Finnish air forces assumed a great deal of responsibility for the air war over their respective fronts. Indeed, on many occasions, Germany's allied air forces conducted missions in support of the German army. At the start of the Russian campaign, the Romanian air force had more than 400 aircraft, the Hungarians about 250 planes, and the Finns about 300. All flew an eclectic mix of their own models, obsolescent license-built aircraft, and foreign aircraft, including British Hurricanes (Romania) and American Brewster Buffaloes (Finland). By all reckoning, these small air forces should not have done well against the large and fairly modern Red Air Force, but the Romanians and Finns in particular performed exceptionally well and destroyed large numbers of Soviet aircraft. Yet, as with Italy, the Luftwaffe showed little interest in building up its allied air forces in 1941–1942.

While the Romanians, Hungarians, and Finns made huge sacrifices and committed their whole armed forces to fight alongside the Germans, the Germans showed them little regard initially. Finnish pilots fought brilliantly against the Soviets in 1939–1940, despite being grossly outnumbered. If any air force could be expected to use modern equipment to maximum advantage, it was the Finns. Yet the only support the Germans were willing to give them in 1941–1942 was to sell them some captured Curtiss Hawk and some ex-French air force Morane-Saulnier MS 406 fighters—the latter being considered a thoroughly mediocre aircraft even by 1940 standards.

By mid-1942, the Germans began to realize that it was not going to be a short campaign in Russia. The mix of obsolete aircraft flown by the Finns, Romanians, and Hungarians was no match for the new aircraft of a resurgent Red Air Force, so the Luftwaffe began supplying its allies with some modern aircraft, including Me 109 fighters. Still, Germany's allies had mostly obsolete aircraft. It took the disaster at Stalingrad to change the Reich's policy on licensing German technology and encouraging its allies to expand their airplane manufacturing. In November 1942 the Luftwaffe finally granted the Romanians a license to manufacture the Daimler Benz DB 605 engine that powered the Me 109G, and in 1943 it agreed to a scheme in which Me 109s would be built in Romania. Small-scale license production of the Me 109 began in Hungary in late 1942. The Germans developed grand plans to build large aircraft factories in Hungary, which would have started mass production in late 1944 if the Soviets had not overrun most of the country.

Given their limited finances and technical resources, Germany's allies produced a surprising number of aircraft. Total production of Hungarian-designed and licensed foreign aircraft during the war was 1,556. Romania produced 450 of its own IAR 80 fighter planes and several hundred other aircraft between 1939 and 1944. Italy manufactured 10,000 aircraft between 1939 and 1943; this is an impressive total, considering Italy's financial straits and technical problems. Given even moderate financial and technical support at the start of the war and licensing for its first-line aircraft, Germany's allies could have become major centers of aircraft production by 1941—just as the Luftwaffe was starting to feel the effects of the war of attrition. With German help, Italy could have doubled or tripled its aircraft production and built first-rate aircraft instead of the obsolete designs it produced. A more powerful Italian air force in 1941–1942 would have changed the whole complexion of the war in the Mediterranean.

An example of what the Italians, Romanians, and Hungarians might have accomplished with German support is found in the case of Canada. In 1939 Canada had a minuscule aircraft industry, producing only sixty light aircraft that year. Starting in 1939, American and British money and technical assistance flowed into Canada to create an aircraft industry from scratch. By 1942, Canada was a major aircraft producer for the Allies, manufacturing 3,622 aircraft that year—more than in Italy. In addition to trainers, fighters, and dive-bombers, Canadian aircraft firms were able to produce more than 2,000 Lancaster heavy bombers for the RAF, one of the largest and most complex aircraft of the war.

Training—The Luftwaffe's Greatest Failure

The Luftwaffe began the war with well-trained pilots. In 1940–1941 German pilots went into operational units with approximately 250 hours of total flight time, about 100 hours of which were in the aircraft they would fly into combat. This compared favorably with the RAF, which sent pilots into action with about 200 hours of training and only 60 to 75 hours in operational aircraft.

It takes eighteen to twenty-four months and a few hundred flight hours to train a fighter or bomber pilot to the point of effectiveness in combat. At the start of the war, Britain realized that airpower would play a key role in the war and that many thousands of pilots, navigators, bomb aimers, and other aircrew would be required. A vast system capable of training tens of thousands of pilots was set up for British Commonwealth aircrews at Canadian bases. Starting in 1940, the United States also began construction of a vast training establishment. Although German pilots held the training edge at the start of the war, it did not last long.

While the Allies embarked on a vast aircrew training program, the Luftwaffe, impressed by the relative ease of its early victories and confident that Britain and Russia would soon fall, carried out only a small expansion of its training program in 1940–1941. Yet even this was crippled by the Luftwaffe's policy of stripping the training schools of experienced instructor pilots and personnel at the start of each major campaign. Since the Luftwaffe had no reserves of pilots or aircraft, the training schools served as the reserve, and training suffered accordingly. Because it was assumed that each campaign would be short, the Luftwaffe leadership easily accepted what was expected to be a short-term disruption.

By 1941, the Luftwaffe was already suffering from a shortage of trained pilots, so the training curriculum was reviewed to see whether pilots could be trained faster and with fewer flight hours. By the second half of 1942, the total flying time for German pilot training had dropped to under 200 hours. In contrast, the British and American training programs had increased their flying time to over 300 flight hours for each new pilot. From then on, the training disparity only grew wider. When the German pilot shortage became acute in 1943, every shortcut was taken to abbreviate training and replace the steady losses suffered not only in Russia but also from the massive air offensive by the Western Allies in the Mediterranean and over Germany itself. By mid-1943, new German pilots flew into battle with only 150 flight hours and a mere 25 hours in their operational aircraft. At the height of the battle for air superiority over Germany in early 1944, Luftwaffe fighter pilots were being thrown into action with only 100 total hours of flying time. They faced American fighter pilots who entered combat with 325 to 400 hours of flight training, including 125 to 200 hours in operational aircraft.

In air-to-air combat, even slight advantages in aircraft firepower, turn radius, climbing ability, and diving speed can be decisive if the pilots are relatively evenly matched. But this was not the case in the skies over Germany. The Me 109s and FW 190s defending German skies were fine machines but inferior in most respects to the American P-51s and P-47s and the RAF's late-model Spitfires. The highly experienced aces of the Luftwaffe knew how to use their planes to the best advantage and continued to shoot down a large number of Allied bombers and escorting fighters. But there were not many experienced fighter pilots left in the Luftwaffe by early 1944. The replacement pilots who filled the ranks of fighter wings were so poorly trained that they did not need Allied fighters to bring them down in large numbers. Inexperienced pilots could not cope with bad weather, engine trouble, or landing on rough airfields. From January to May 1944, the Luftwaffe lost more planes and men to operational accidents than in combat. New German pilots who flew against the highly trained American and British pilots in 1944 were unlikely to last more than a few weeks in combat before being killed or wounded.

Even when the Germans could field clearly superior aircraft, the collapse of the training program prohibited their effective use. One example is the Me 262 jet fighter, a "wonder weapon" that could outfly the best Allied fighters by a wide margin. By the summer of 1944, mass production of the Me 262 had begun, and some were available to attack the Allied beachhead in Normandy. A wing or two of Me 262s could have threatened the Allied air superiority over Normandy, with enormous consequences for the ground war. Yet the Luftwaffe could not use the superior technology of the Me 262 to advantage because hardly any pilots had been trained to fly it. In July 1944 nine Me 262s were ordered to France. At the time, the unit commander rated only four of the nine pilots as being even moderately competent to fly the plane. Due to constant engine problems, the Me 262s accomplished little in France and were withdrawn. The Luftwaffe was able to create and train one group to fly the jet against American and British bombers, but it was not ready for battle until 1945—long after any hope of regaining air superiority had passed.

Conclusion

The primary cause of all the Luftwaffe's failings was bad leadership. That Hitler was no strategist is well known. He had an early run of success when he correctly gauged the weakness of the Western democracies and their reluctance to fight over the Rhineland, the Austrian *Anschluss,* and finally the annexation of Czechoslovakia. He was surprised when Britain and France went to war over Poland, but German rearmament was well ahead of the democracies, and Germany handily defeated France and ran the British army off the Continent. With a hubris born of easy, early victories, Hitler saw no need to mobilize the German economy for the war effort, even as his enemies assembled their full strength for the counterattack.

Hitler's optimism and his contempt for Germany's enemies infected the top leadership of the Luftwaffe. The vain and ambitious Göring shared Hitler's contempt for the United States and Britain, and he failed to take the industrial potential of those nations into account. Göring was also responsible for the mismanagement of the aircraft industry. Although he had been a fighter ace in World War I, he had little knowledge of modern aviation when he took over as commander of the Luftwaffe in 1935, and he made little effort to learn anything about new technologies or aircraft production techniques. In putting his old flying comrade Udet in charge of aircraft production, he could not have made a worse choice. Udet's mismanagement and poor procurement decisions between 1936 and 1941 ensured that Germany fell so far behind in aircraft development and production that the Luftwaffe could never recover.

The Luftwaffe chief of staff from 1939 to 1943, Colonel General Hans Jeschonnek, also played a central role in the decline of the Luftwaffe. Jeschonnek came to the job with a brilliant reputation that was not upheld by his performance. He never understood the real duties of a chief of staff: doctrine, training, building units, procuring equipment, and overseeing logistics. He was fundamentally uninterested in training and logistics, and both of those vital aspects of modern air war declined under his tenure. Jeschonnek devoted almost all his time to operational planning and command—preferring to direct the efforts of the field commanders rather than take responsibility for more mundane but equally essential tasks. The failure of the Luftwaffe to supply Stalingrad in early 1943 was a severe blow to him. In August of that year, the RAF was able to level a large part of Hamburg, killing tens of thousands of German civilians while suffering few losses to Luftwaffe defenses. Having failed the nation and Hitler so dramatically, Jeschonnek committed suicide. His successor was the very competent General Gunther Korten, who came on the scene too late to arrest the decline of the Luftwaffe. At the top levels of command, only Field Marshal Ehrhard Milch, state secretary for aviation and responsible for many of the day-to-day operations of the Luftwaffe, had a sound understanding of aircraft production and training requirements. However, his warnings were usually ignored until the middle of the war.

The Luftwaffe was crippled by a flawed short-war strategy that infected the whole Nazi high command. The Germans were so sure of an early victory that they ignored the requirements of a war of attrition. The short-war dogma hindered industrial mobilization and expansion of the training system and precluded real assistance to allied nations. Germany was stuck in this ruinous outlook until the disaster at Stalingrad forced even Hitler to accept that it was a war of attrition. By that time, the Allies had won an insurmountable lead in the air war.

Suggestions for Further Research

The Luftwaffe offers plenty of material for further study, but because its archives were destroyed at the end of World War II, it is hard and painstaking work to reconstruct detailed operational histories. Although some campaigns, such as the 1940 Battle of Britain, have been thoroughly explored, for other major campaigns, the solid historical work is pretty thin. Some areas in which further research and writing would be welcome include the Luftwaffe in the Mediterranean campaigns (1942–1945) and the Luftwaffe's failed campaign against the Allied landings in Normandy. One of the most interesting technological developments of World War II was the employment of the first true precision bombs—radio-guided bombs dropped by the Luftwaffe against Allied ships in the Mediterranean in September 1943. This story has not been told in detail.

The largest gap in the literature on the Luftwaffe is in biographies of senior leaders. There are numerous books and memoirs of Luftwaffe fighter aces, but very little has been written about the senior commanders who led the Luftwaffe against the combined might of Britain, the United States, and the Soviet Union. David Irving has written biographies of Milch and Göring. Both contain some useful information, but this is often marred by some of the author's absurd assertions. For example, he argues that Göring did not know of the Holocaust.

Some highly capable senior officers, such as Field Marshals Hugo Sperrle and Wolfram von Richthofen and General Alexander Lohr, led the Luftwaffe air fleets. Little has been written about these men, and all deserve full biographies. The Luftwaffe's chiefs of staff, Jeschonnek and Korten, both deserve proper biographical attention.

Recommended Reading

The best and most thorough coverage of Luftwaffe operations in World War II is found throughout the ten-volume German official history of the war written and published by the staff of the Bundeswehr Military History Research Office. The extensive sections on all aspects of the Luftwaffe were written and edited by first-rate airpower historians such as Klaus Maier and Horst Boog. Most of the German official history has been translated into English (up to volume 7) and is published by Oxford University Press under the title *Germany and the Second World War.* The best general work on the Luftwaffe's leadership, training, doctrine, and organization is Horst Boog's *Die deutsche Luftwaffenführung 1935–1945* (Stuttgart: Deutsche Verlags-Anstalt, 1982). Unfortunately, this highly detailed and wonderfully researched work has not been translated into English.

On the evolution and development of the German aviation industry and the Luftwaffe's prewar buildup, Edward Homze's *Arming the Luftwaffe* (Lincoln: University of Nebraska Press, 1976) has long been considered the best work. My book *The Luftwaffe, Creating the Operational Air War, 1918–1940* (Lawrence: University Press of Kansas, 1997) covers the development of the Luftwaffe's doctrine and organization from the interwar period to the early years of World War II in some detail. The most useful book in English on the Luftwaffe's doctrine is James Corum and Richard Muller, *The Luftwaffe's Way of War* (Baltimore: Nautical and Aviation Press, 1998). The authors have translated and edited the most important air-war doctrine documents written between 1910 and 1945. The collection is quite comprehensive and provides those who cannot read German a clear picture of German airpower thinking in one volume.

Williamson Murray's *Strategy for Defeat: The Luftwaffe, 1933–1945* (Maxwell Air Force Base, Ala.: Air University Press, 1983) is a good general history. Its strong point is its coverage of the Luftwaffe's aircraft production, but it is fairly thin on the history of Luftwaffe operations. Mathew Cooper's *The German Air Force, 1922–1945: An Anatomy of Failure* (London: Janes, 1981) is a good general history but is out of date in parts.

Several very good histories of Luftwaffe operations have appeared in the last decade. E. R. Hooten, *Phoenix Triumphant* (London: Arms and Armour 1994), is a very readable and well-researched overview of the Luftwaffe's interwar years and the early campaigns of World War II. One of the most masterful works on Luftwaffe operations is Joel Hayward's *Stopped at Stalingrad: The Luftwaffe and Hitler's Campaign in the East, 1942–1943* (Lawrence: University Press of Kansas, 1998). Hayward's work covers the air war in Russia during the decisive campaign of 1942 and places the major air operations into the broader strategic context. Another good operational history is Adam Claasen's *Hitler's Northern War: The Luftwaffe's Ill-fated Campaign, 1940–1945* (Lawrence: University Press of Kansas, 2001). Claasen's book covers the Luftwaffe's air operations in Norway from the German invasion in 1940 to the air strikes against the Allied convoys to Russia. Until recently, this important theater of war had been ignored.

Certainly, the most vital air campaign of the war was the Luftwaffe's defense of Germany against the Allied bombing offensive. If the Allies had not defeated the Luftwaffe and won air superiority, the 1944 landings in Normandy could not have taken place. A superb account of the Luftwaffe's fighter force is found in Donald Caldwell, *JG 26: Top Guns of the Luftwaffe* (New York: Orion Books, 1991). Caldwell depicts the meteoric rise and then slow decline of the Luftwaffe's fighter force through the lens of the fighter wing that bore the brunt of the defense in the west from 1941 to 1944. One of the finest authors on the Luftwaffe is Alfred Price, who has a thorough understanding of the relevant technology and equipment; he provides the nonspecialist reader with a clear description of electronic warfare, Luftwaffe aircraft, and the advantages that even minor adaptations in design and technology gave to one side or the other. Two of Price's most useful books are *Luftwaffe Handbook, 1939–1945* (New York: Charles Scribner's Sons, 1977) and *The Last Year of the Luftwaffe* (London: W. J. Williams and Son, 1991). The important story of the German night fighter units is told by Peter Hinchliffe, *The Other Battle: Luftwaffe Night Aces versus Bomber Command* (Edison, N.J.: Castle Books 1996).

One extremely important part of the Luftwaffe and the air defense of Germany was the flak forces, which made up about 40 percent of the Luftwaffe's total personnel. About half of the Allied aircraft lost over Germany were lost to the flak guns, but the flak force has never received even a fraction of the attention given to the more glamorous fighter aces. In *Flak: German Anti-Aircraft Defenses, 1914–1945* (Lawrence: University Press of Kansas, 2001), Edward Westermann provides a thorough and readable account of the development and employment of the Luftwaffe's flak forces, bringing a broader perspective to the reader's understanding of the air war over Europe.

For an overview of the air war, see Richard J. Overy, *The Air War, 1939–1945* (New York: Stein and Day, 1980), and *Why the Allies Won* (New York: W. W. Norton, 1995).

The Japanese left its vital technical ground staff in unhealthy conditions in the Southwest Pacific islands and then abandoned them at airfields like this one in New Guinea.

The feared Zero fighter is pictured at a Flying Tiger base in China with a P-40. The Zero was another prewar decision that failed to keep up with the demands of modern war.

A Zero sandwiched between its opponents (a P-40 above and a P-51 below) over China.

The Luftwaffe was expecting a short war, so the Me 109G was still a frontline fighter in 1945, even though its poor engine design and flimsy undercarriage were hazards operationally. This postwar photo shows a fighter in the Spanish fleet.

The versatile German Ju 88 was used as a night fighter in defense of the Third Reich.

Hitler was always interested in novel war-winning ideas, such as the Me 263 twin jet and the rocket Me 163. Flying the latter was hazardous to one's health.

Russian soldiers on field maneuvers prior to 1914, when the French-designed Farmans were built under license.

By 1934, the Soviet air force was the largest in the world just before Stalin's purges. Here a Maxim-Gorky drops parachutists.

The Polikarpov I-16 was an extremely nimble fighter with which Russia entered World War II. Lack of oxygen equipment did not matter, as most operational flying was at less than 3,000 feet.

The Japanese were very conscious of the importance of fuel and refueling equipment, as these wrecked tankers at Bellows Field, Hawaii, show.

Lieutenant Commander Minoru Genda, the Japanese assistant naval attaché in London, traveled across North America on his way home with details of the 11 November 1940 British raid against the Italian fleet at Taranto.

Some of the defending fighters in the Philippines were Boeing P-26 "Peashooters," each with its own pilot and crew chief.

A combat formation of Boeing P26-A airplanes.

The early campaigns of World War II were fought with the machines and ideas of the early to mid-1930s. The Gloster Gladiator was the last in a long line of interwar biplane fighters. It was still on home and overseas squadrons into 1941.

The Lysander had speed and short takeoff and landing capabilities, but it was too large for the army cooperation role.

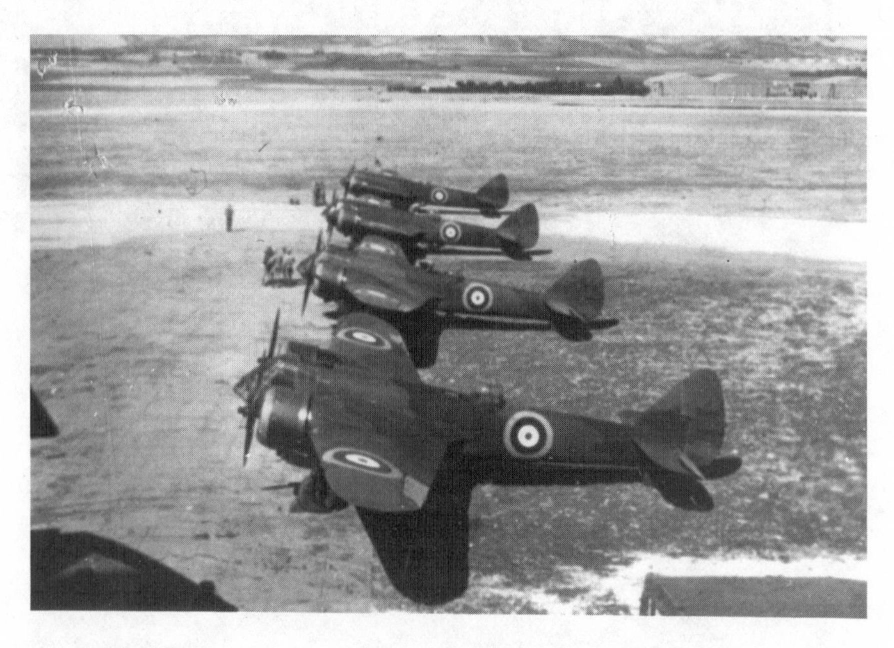

The Blenheim I was a fast medium bomber with insufficient defensive armament and tactics, based on its performance and characteristics. Even so, it was sent on raids into Germany as late as May 1942.

CHAPTER EIGHT

The Argentine Air Force versus Britain in the Falkland Islands, 1982

René De La Pedraja

The defeat Argentina suffered in the 1982 war over the Falkland Islands (Islas Malvinas) was so disastrous that there is a widespread perception that it never stood a chance against Britain. An image of overwhelming British superiority emerges whether reading British authors praising their own forces or Argentine writers trying to account for the inevitable defeat.

The efficiency, audacity, and, in particular, determination of the British forces are beyond question. But if British victory was such a foregone conclusion, the student of military history has little to learn from this war. A detached examination of the operations, however, reveals that the outcome was in doubt until the last moment and thus is useful to identify the factors that finally tipped the scale in favor of the British.

The assumption in this chapter is that Argentina stood a very good chance of winning the war. The military junta then ruling Argentina committed many blunders, but its conclusion that Britain was vulnerable was an accurate assessment. As will be shown, the real explanation for the defeat of Argentina lay in the junta's inability to employ its military forces adequately and at the right moment. The truth of this proposition is easily proven in many ways. For example, if the junta had waited until 1983, when Britain planned to dispose of its two aircraft carriers, then by default, the Argentine navy would have achieved a superiority that precluded any British reconquest of the islands. Or if the junta had followed the original plan to launch the invasion at the later date of 15 May 1982, there would have been no time for a serious British response that year before the start of winter. Instead, the early start of the war on 2 April opened a window of opportunity for Britain to complete the offensive before serious winter weather set in.

In the simplest scenario, if the army, navy, and air force had worked effectively and harmoniously, Argentina could have won the war easily. If just two of the services had functioned properly, Argentina would have won after a hard struggle. But if only one service was effective, victory would depend on an extremely high level of performance against the British enemy. Of the three services, first the navy and then the army largely dropped out of the war. The air force desperately tried to fill the gap, and at times its heroic performance was on the verge of saving the other two services, but ultimately the navy and the army dragged the air force down into defeat. That the air force came so close to winning the war for Argentina is a renewed testament to the impressive impact of airpower in modern warfare. By examining the near victory of the Argentine air force, this chapter offers many lessons on the potential of airpower.

The Air Force in the Prewar Military Environment

Until the economic meltdown of 2001, Argentina enjoyed the reputation of being the richest country in Latin America. Possessing perhaps the most fertile soils in the world, Argentina was a breadbasket that traditionally exported high-quality foodstuffs at low prices. In the early 1900s Argentina ranked among the ten richest countries in the world. Although primarily an agricultural exporter, a strong industrial base emerged in this highly urbanized country by the 1960s. European immigrants or their descendants constituted the majority of the population. Although second largest in size among the Latin American countries, Argentina was sparsely populated. A nationwide peasantry did not exist, and the well-educated urban population seemed more than ready to take the country into the next stage of its economic growth. Argentina was the most European of all the Latin American countries. Buenos Aires, the capital, had such an intense cultural and intellectual life that many referred to it as the Paris of the region.

The international prestige enjoyed in the early 1900s began to vanish by the middle of the twentieth century as Argentina slipped from the ranks of rich countries. Because its neighbors suffered greater deterioration, however, Argentina retained its image as the richest country in Latin America. In reality, considering it the least poor country in the region was a more accurate depiction. Beginning with the Great Depression of 1930, Argentina suffered a long period of decline, first in the economy and then in other areas of human activity. Periods of recovery, such as the World War II boom, only partially reversed the downward trend. Equally revealing, early mistakes turned into a consistent pattern of errors, failures, and even blunders after 1930. Whether in the economic, political, social, or diplomatic sphere, the leaders of the country inexorably made decisions that worsened the decline and minimized the recoveries. Within the broad debate of whether military defeat occurs independently of other trends in a society, the military collapse of Argentina in 1982 fit seamlessly into the pattern of repeated failures during the twentieth century.

After the destruction of the last Indian tribes to the south in the nineteenth century, the threat of war with Chile and Brazil remained Argentina's primary defense concerns. Large armies were indispensable for any land war with its neighbors, and to meet the need, the Argentine congress approved a law creating a conscript army in 1901. At least until World War I, this army seemed formidable and possessed the best weapons available, such as Krupp cannon and Mauser rifles. The government did not want to assume the high cost of inducting all eligible males into the armed forces and selected a more modest alternative. According to the strategic needs of the army and navy, the required number of conscripts was picked by lottery. In practice, only a small percentage (usually around 10 percent) of the eligible males served in the military. The term of two years in the original law declined to one year for the army and fourteen months for the navy. Apparently, during the first half of the twentieth century, both the army and the conscripts gained something from the experience. New imported equipment and adequate funds made it possible to train many of the conscripts, who acquired habits of discipline and organization that would be useful later in life, and the army succeeded in creating reasonably competent combat units. But after World War II, declining budgets and obsolete equipment made the conscripts' year in the army a waste of time. A growing number of legal exemptions fostered an environment of favoritism and outright corruption. The draft became an obstacle that many young men tried to dodge; in practice, those without money or influence ended up in the army. At best, the conscripts spent their year of military service loitering in the barracks; at worst, brutal officers exploited the conscripts as menial laborers.

Like most other air forces, that of Argentina emerged from the army. In 1919 the army created a separate bureau for aviation, but it was 1935 before aviation acquired the status of a separate branch of the army, equivalent to the artillery, cavalry, and infantry. A series of changes finally gave the air force formal independence from the army in 1945, but by then, the air force had assimilated many army practices. Most notably, the air force adopted the army's conscript model and did not attempt to operate under a system of purely voluntary recruitment. For example, in 1980 and 1981, slightly more than half the membership of the air force consisted of conscripts (10,000 of 19,500), but unlike army draftees, those in the air force performed their duties competently and even admirably. Most of the capable and dedicated conscripts insisted on spending their year of duty in the air force. Many saw the air force as an opportunity to advance their careers by learning manual skills and intellectual abilities that would be useful in civilian life. In contrast, the army was interested merely in filling its recruitment quotas and did not seem concerned about losing the best conscripts to the air force (and, to a lesser degree, the navy).

Not surprisingly, the reenlistment rate in the air force was much higher than that in the army. In addition, a steady stream of volunteers provided most of the pilots. Aviation had touched a sensitive nerve in Latin America, and many persons, whether in Colombia, Mexico, or Argentina, looked to the skies with great expectations. Whether for aristocrats seeking meaning in their often empty lives or poor persons struggling for upward mobility, aviation generated a fascination that attracted many to the skies. Argentina enjoyed an abundance of dedicated and intelligent men who wanted to be pilots. They longed to push their planes to the limit and wanted to experience risk and excitement.

As in countries in more advanced stages of industrialization, the Argentine air force operated extensive facilities and arsenals for the production of airplanes and aeronautical equipment. Argentina produced a variety of piston engine planes but was unable to break into the market for jet planes after World War II. Because of capital and technology requirements, the entry costs into the production of jet engines proved prohibitive. Argentina could repair but could not build jet warplanes.

The air force realized that it had more pilots and skilled personnel than it could reasonably employ, even in the pre-1950 period of moderate budgets. In a prescient move that later helped cushion the adverse effects of declining funds, the air force assumed many duties usually performed by civilian agencies in other countries. For instance, it regulated and controlled all aspects of civil aviation, such as airplane registry and airport operation; even the national weather service was under its jurisdiction. As is often the case in industrializing countries, the air force operated its own airline—Lineas Aéreas del Estado (LADE)—which flew to remote areas, most notably to Patagonia, bordering the Falkland Islands. Superficial critics considered these activities a distraction from the true military duties of the air force, but in reality, these civilian functions were a key to its extraordinary performance. Not only did the pilots build up flying hours, but they and the ground crews also learned to operate a wide range of airplanes in the harsh, windy, freezing environment of the south. Because many of the planes lacked the design or the instrumentation necessary for flights to the south, the pilots and ground crews became unusually adept at keeping the planes flying under extremely adverse conditions. The dexterity of the personnel kept accidents to a minimum during the war, in spite of the bad weather. For example, few accidents occurred during the landing of planes in the Falkland Islands. The experience of handling air traffic in Patagonia brought unexpected benefits when air force personnel efficiently operated the airport at Port Stanley during the war.

As the air force strove to cope with inadequate budgets, its strategic outlook underwent two major changes in the second half of the twentieth century. Although Chile remained a potential enemy on land, Brazil had ceased to be a likely opponent by the 1930s. Fear and rivalry gave way to strong business, trade, and personal links with Brazil. Thus, Argentina no longer faced a land threat from Brazil, so any danger of a two-front war had vanished. Only the boundary dispute with Chile remained a nagging worry, and Argentina's overwhelming superiority in resources, area, and population made it unnecessary to maintain a large conscript army. By the 1950s, Argentina needed to phase out its conscripts and concentrate its resources on a volunteer and highly professional military. Advanced technology would have made the air force a prime candidate for expansion under the new volunteer model, had not a second strategic change driven the air force in a strange direction.

After World War II, Argentina, along with the rest of Latin America, found itself dragged into the Cold War, a struggle for global hegemony between the United States and the Soviet Union (today's Russia). After initial resistance to the pretensions of the United States, the Argentine army gradually adopted the Cold War, or the supposed struggle against communism, as a way to justify the continued existence of a large conscript army. A series of internal meetings, exchanges with the United States, and extensive deliberations produced a change in national strategy between 1955 and 1962. After fierce internal struggles, the army managed to transform the fundamental national strategy from a doctrine of national defense to a doctrine of national security. Although the presence of Chile as a likely enemy did not disappear completely from the new strategy, it dropped to a secondary role. The new enemy became the Argentine people themselves, who numbered in the millions; to stop any subversive drift toward communism, a large army was indispensable to control the population. Under the new anticommunist mission, inept and ill-suited officers could rest safely in the knowledge that their army careers were secure. The usual politicking to promote the worst officers to the most important commands could continue undisturbed.

Of the three services, the air force was the least suited for the new doctrine of national security. It increased its ground security personnel and antiaircraft batteries, similar to what the navy had done with its marines, but not unexpectedly, the army repeatedly blocked any large expansion of air force ground combat units. To retain a meaningful role under the new doctrine, the air force found a more promising avenue in counterinsurgency. Although counterinsurgency was a basic component of the doctrine of national security, a role for the air force was a hard sell in Argentina. The absence of jungles or peasants made any rural guerrilla movement impossible. Only in the subtropical north, near Paraguay, could thick forests hide guerrillas, but the extremely sparse population in those regions presented an insurmountable barrier to guerrilla recruiters. The overwhelming urban population made city-based guerrillas and terrorists the most likely opponents, as was the case for a few brief years. Any possible air force contribution to the campaign against urban guerrillas seemed minimal at best. Nevertheless, counterinsurgency was the policy the air force had to adopt to survive in the new hostile environment.

Because its planes were inappropriate for counterinsurgency warfare, air force factories began to develop a new model, the Pucará, an airplane suitable for a role that was largely nonexistent in Argentina. The twin-engine turboprop was a modest commercial success, however, as other countries saw this plane as an answer to their counterinsurgency challenges. At a time when Argentina had the capital and the talents to produce at most two models of subsonic planes, the decision to develop the Pucará was questionable at best. Preliminary analysis suggests that the Pucará diverted effort, time, and resources from other types of aircraft that would have been much more helpful in a conventional war. After the Falklands war the air force factory produced the Pampa, a popular jet trainer that has satisfied both military and profit considerations. Admittedly, Argentina lacked the industrial base to produce supersonic jet fighters and electronic components for modern aerial warfare, but the Pucará still seems an odd fit. A review of the aerial operations during the war reveals that the development of a fast long-range plane for aerial reconnaissance would have been a wiser choice.

A long-range reconnaissance plane equipped with maritime navigation and surveillance electronics was within the technological capabilities and capital resources of the air force but was simply out of the question because of the turf war raging among the three services. In a historic and momentous decision in 1969, the navy received exclusive jurisdiction to defend Argentina from sea attack. Even more incredibly, this decision specifically prohibited the air force from developing or acquiring any capability suitable for aerial operations over the sea. Only at the request of the navy could the air force play a secondary or supporting role. This 1969 decision effectively trapped the air force inside the continental landmass. Even a cursory examination shows how irrational and incongruous this allocation of tasks was. The northern third of the seacoast contained the main bases of the Argentine navy, while Patagonia in the southern third of the country hosted the landing strips and airports established by the air force and its subsidiary LADE. Nobody was planning for a war over the Falklands then, but already in the 1970s the air force, with its preliminary deployment in Patagonia, was the service best positioned to carry out any offensive maneuver. Had the government allowed the air force to acquire, equip, and train a fleet of airplanes suitable for combat over the ocean, the outcome in 1982 would have been much different. "But the time wasted since 1969 could not be recovered," as Pío Matassi ruefully commented in 1994.

At first glance, the navy might have seemed the proper choice for sea defense, had not a number of factors undermined its natural role. As mentioned, the bases and forces of the navy were concentrated primarily in the northern third of the seacoast, particularly near Buenos Aires, to allow the admirals the maximum opportunity to participate in the turbulent politics of the country. The navy had been the premier service in the first half of the twentieth century, and although some units, such as the marines, maintained that tradition, the quality of the officers and the preparedness of many ships had declined, sharply in some cases. Argentine admirals, perhaps inevitably for any navy, had placed the highest priority and spent most of their budgets on ships, particularly surface vessels. However, the admirals did not have the same sense of urgency about submarines and airplanes, the two branches of the navy most likely to inflict crippling losses on the British. In 1982 the navy was in the middle of a fleet renewal program, but deliveries of most new vessels were slated for 1983. The old aircraft carrier *25 de Mayo* appeared to give the navy a formidable striking power at long distances, but in reality, the fleet wing consisted of eleven A-4 Skyhawks, only eight of which were still operational when the war began. To replace the aging Skyhawks aboard the small carrier, the navy had purchased fourteen Super-Étendard planes with twenty Exocet air-to-ship missiles. Crews were training in France with the new planes, and the first shipment of five Super-Étendards and five Exocets had reached the navy only in November 1981. The second lot of five planes and five Exocets was scheduled to arrive in April 1982. British authorities have tacitly implied that the presence of twenty or possibly just ten of the missiles would have required them to abort the plan to send a naval task force. Instead, in April 1982 the Argentine navy could offer only a small old carrier, eight Skyhawks, and five Super-Étendards with five Exocet missiles for the defense of the islands.

Obviously, the modest naval aviation arm could not stop the British navy on its own, so in spite of the prohibition on its participation in sea operations, Argentina inevitably had to turn to its air force. What no one could foresee was that the air force would end up carrying the brunt of the war. At first glance, the air force's arsenal seemed formidable. It had more than 200 combat planes, including 19 French Mirage IIIEA fighters, 26 Israeli Dagger fighter-bombers, and 9 British Canberra bombers. The A-4 Skyhawks were the largest component and may have numbered as many as 68; there were also at least 45 Pucarás. A variety of training and transport planes, as well as helicopters, rounded out the Argentine air force. But the seemingly powerful service had severe limitations, as the specifications in the equipment manuals clearly revealed. Of all its combat planes, only the Canberra bomber had the range to fly to the Falkland Islands and back, but it was the model most vulnerable to missile and Harrier attack. The prize planes, the Mirages and Daggers, could reach the islands only if they sharply curtailed their flying time at supersonic speeds; worse, aerial refueling was impossible. The refuelable Skyhawks could reach the islands, but not if they were carrying full bomb loads; even then, they could not tarry more than a few minutes. The specifications of the airplanes confirmed that Argentina had acquired an air force suitable for short-range ground support missions. A war with Chile had been the expected scenario, with the planes flying from the many airfields near the long land border with its western neighbor.

In 1978 the previous military junta had mobilized a quarter of a million men for an invasion of Chile. A last-minute intervention by the pope in December 1978 had averted the conflict but did not end the danger of hostilities in the future. The best option for Argentina was peace with both Chile and Britain, and the worst option was a two-front war with these two countries. In the case of a single-front war, which war was the best option for Argentina? With the advantage of hindsight, and knowing the outcome of the Falklands war, it is clear that Argentina should have attacked Chile in 1979. A victory for Argentina would have been highly likely but not inevitable; an embarrassing stalemate might have ensued, but Argentina's preponderance of resources eventually would have brought victory, and the many flaws in the Argentine military performance would have been exposed. After correcting those defects, Argentina would have been in a better position to challenge Britain in 1982. Alternatively, a costly war or a bloody stalemate with Chile would have persuaded Argentina to keep the dispute with Britain within diplomatic channels.

The many problems and the high cost of the mobilization in 1978 were well known throughout the military, and in some cases, reports even reached the public through newspaper accounts. The new Galtieri military junta, which replaced the previous junta in December 1981, was composed of members who had participated in the 1978 Chilean mobilization. Apparently, however, they had been isolated from reality and did not understand the seriousness of the deficiencies. Thus, even before taking office, the three members of the new military junta—Leopoldo Galtieri of the army, Jorge Anaya of the navy, and Basilio Lami Dozo of the air force—displayed a strange detachment from events and had tremendous difficulty keeping in touch with reality. Within the new junta, the air force was in the weakest political position. To become the new president of Argentina, Galtieri had secured the vote of his childhood friend and fellow junta member, Admiral Anaya. Facing a majority of two in the three-man junta, the air force chief could only acquiesce. As part of the price for securing navy support to become president, Anaya had obtained a promise for the eventual occupation of the Falkland Islands from a willing Galtieri. Only a month later, in January 1982, did these two junta members let Lami Dozo in on the plan. Because the navy and army planned to carry out the operation with only their own resources, the air force could not block the occupation idea. Furthermore, the initial plan was vague about dates, and the proposed operation seemed to lack urgency. In any case, the air force did nothing prior to 26 March 1982 to prepare for a war in the Falklands—an understandable posture, given that it was legally forbidden to prepare for maritime operations.

Occupation of the Islands and Preparations to Repulse the British

The original plan had called for an occupation of the islands on 15 May, and in no case earlier than 15 April. As mentioned previously, those dates would have deprived the British of enough good-weather days to recapture the islands in 1982. Unfortunately for Argentina, in late March a diplomatic dispute arose over whether British authorities had properly authorized Argentine scrap-metal workers to dismantle rusting whaling installations in the remote island of South Georgia. A prompt diffusion of this incident involving the Antarctic-like and virtually deserted island was indispensable to keep Argentina's invasion plans secret. Instead, Admiral Anaya seized on the South Georgia episode as a pretext to press for the occupation of the Falklands and, more significantly, for an earlier invasion date—2 April. It is easy to put all the blame on the military junta for starting the war, but the civilians in the Foreign Ministry, right up to the blindly inept foreign minister, failed to sound even the most elemental warnings about the certainty of strong hostility from both Britain and the United States. The military government had such an atrocious reputation in the world that even the United States had put an embargo on new arms sales. The national security doctrine had resulted in the torturing and killing of thousands of innocent civilians, but the Foreign Ministry failed to convey any sense of the international outrage at these human rights violations. All the ministry had to do was submit a sample of the countless hostile articles in the world press, which the junta obviously did not read. According to the Rattenbach Commission, "Under these conditions the rush to occupy was inexplicable. In an obsession to preserve a strategic surprise, the junta picked the worst possible moment from the perspective of foreign relations" (25).

An overwhelming consensus exists that the junta's decision to invade the Falklands in response to the South Georgia crisis was a disastrous blunder. However, in itself, the decision to invade was not flawed. It was the timing—specifically, moving the invasion date up to 2 April—that proved fatal. The dynamics of junta politics explain why the air force could not stop the mad rush to invade. As the politically weakest member of the junta, Lami Dozo was well aware that he served at the pleasure and the sufferance of the other two members. It was not just a question of replacing Lami Dozo with a more pliant general; rather, the very membership of the air force on the junta was at stake. The air force had feet of clay. Although the army garrisons near most air installations supposedly offered protection from attack, in reality, the bases ensured the good political behavior of the air force; thus, seizure by army troops hung as a permanent threat over the air force. Paradoxically, the strongest service militarily was the weakest politically when it came to the junta's decisions about war and peace. Because the invasion was a joint army–navy operation, Lami Dozo was not even in the chain of command, so he could do nothing to modify, much less delay, the operation. He could only support the decision enthusiastically if he wanted to remain a team player. He was fully aware that the navy and army wanted to take all the credit for the recovery of the islands.

Only President Galtieri could stop or at least delay the invasion, but he too depended on navy support to guarantee his position. On military grounds alone, he should have insisted on delaying the invasion at least until late April. In Lawrence Freedman and Virginia Gamba-Stonehouse's opinion, "The start of April 1982 was about the last moment that Argentina could have chosen to allow Britain to mount a counter-response" (98). But Galtieri could stop the invasion only if he was ready to sacrifice his presidency for the sake of the country. As a team player who had risen to his high leadership position by not antagonizing anyone, it would have been out of character for him to defy the growing momentum behind the demands of Admiral Anaya. On Friday, 26 March, the junta made the decision to occupy the Falklands. U.S. intelligence was lax in detecting the telltale signs, and by the time President Ronald Reagan telephoned with a personal appeal, the deadline to recall the invasion fleet had passed. That phone call should have unambiguously demonstrated to the junta which side the United States intended to support during the war. Nevertheless, with the British fleet engaged in well-reported maneuvers near Gibraltar and thus nearer to the Falklands than the Argentine plans had envisaged, Argentina decided, in Robert L. Scheina's opinion, to carry out the "landing at a time that now looks almost as if it had been picked by Britain."

Special forces flying aboard three air transports would have been enough to capture Port Stanley and would have provided the best means of ensuring secrecy until the occupation, whether it occurred on 2 April or a later date. But because the other services lacked long-distance planes (and could not requisition civilian aircraft for fear of exposing the operation), the exclusion of the air force made a shipborne landing the only alternative. However, nothing required or justified the vast deployment of surface warships that Admiral Anaya insisted on. The use of practically the entire Argentine navy was not only unnecessary but also downright harmful; it promptly gave away the invasion plan. On 26 March the British Admiralty ordered the mobilization of its units as soon as it intercepted Admiral Anaya's orders sending the Argentine warships *Drummond* and *Granville* south immediately. This one week's advance notice, courtesy of the Argentine navy, helps account for the British navy's rapid response to the 2 April occupation. Only an occupation date on or after 15 April could have guaranteed an Argentine victory, but even pushing back the advance notice from 26 March to 2 April would have delayed the British mobilization and thus increased the uncertainty about the outcome of the war.

Practically all Argentine warships, including the aircraft carrier, put out to sea for the invasion of the islands. Nearly a thousand soldiers disembarked to force the surrender of a minuscule British garrison in the early hours of 2 April. The invasion was supposed to jump-start negotiations in the fantasy scenario of the Foreign Ministry; in such a scenario, the invasion force was designed to be too large for a token seizure but too small to defend the islands from the full British fleet. Thus, from the start of the operation, the dimensions were wrong. Rather than warships, merchant ships carrying earthmoving machinery, cement, construction materials, and heavy artillery should have been the first ships to dock. The big fear of the Argentine navy had always been British nuclear submarines; consequently, the navy should have escorted shipload after shipload of supplies, equipment, and weapons before the British tried to blockade the islands, yet the navy did not even press for convoys. A massive and sustained buildup to fortify and supply the islands would have been the best way to demonstrate to the British a dogged determination to keep the territory. Instead, the junta distracted itself with the meaningless gesture of recovering South Georgia; that remote island simply could not be defended, and British warships easily recaptured it on 25 April. The site of the initial incident that sparked the war was thus returned to British control even before April was over.

After the initial Argentine euphoria at the recovery of the Falklands, the junta woke up to the realization that the British were sending almost their entire navy to recover the islands. The junta had not imagined this possibility even in its most extreme scenario. War was not inevitable, however, because at that point—and even later—an evacuation under the cover of diplomatic initiatives could have avoided a humiliating defeat. In a momentous meeting on 4 April, the junta rejected any peaceful evacuation and decided to fight to retain the Falklands. In a surprising reversal, the army and navy, which had previously been so meticulous about excluding the air force, suddenly demanded its utmost participation in the defense of the islands. Over the coming weeks, the air force, by default, assumed the main burden of defending the Falklands.

Such a sudden and dramatic shift in mission caught the air force by surprise. Only on 26 March had air force officials even learned of the invasion order, yet a week later, the navy was presenting an extensive list of specific requests. The air force, which had long been trapped in a ground support role, could not suddenly improvise and carry out the techniques required to attack warships at sea. Furthermore, it had no all-weather aircraft, and its airplanes lacked the range to operate over the islands from mainland bases. Yet these and other reservations did not deter Lami Dozo, who was eager to show the world all that airpower could do.

The can-do tradition is typical of spirited air forces, and Lami Dozo was supremely confident that his subordinates could rise to the challenge on very short notice. On 7 April the General Staff issued a plan of operations, giving the air force less than a month to complete its preparations before the British began hostilities on 1 May. Those weeks were crucial for all three services, but in particular for the air force, which had to try to make up for a lost decade. The more the air force accomplished before that deadline, the greater the probability of defeating the British navy. Most accounts generally take for granted the existence of air bases in Patagonia, but in reality, none were ready for war. The deficiencies were fundamental. For example, some landing strips were too short for combat planes, and existing long runways lacked the access lanes and parking spaces necessary to handle a large number of planes.

The instantaneous cooperation of local citizens and contractors proved indispensable to ready the air bases in Patagonia. In addition to runways, hangars, buildings, and other facilities materialized to allow sustained air operations from Patagonia.

Because of the limited range of the airplanes, Argentina needed to use the airfields closest to the islands. Considerations that might have seemed unimportant at first, such as the location of Port Stanley in the extreme eastern end of the islands, meant additional minutes of flying time. Of the three main air bases in Patagonia, Río Gallegos was about 750 kilometers (469 miles) from the islands, San Julián was 700 kilometers (438 miles) away, and Río Grande at Tierra del Fuego was 690 kilometers (431 miles) away. Partly for reasons of dispersal, and partly because of the inability to handle more flights, the air force scattered its planes among the three airstrips rather than concentrating them at Río Grande, the base nearest to the combat theater. (The long range of the Canberras allowed the air force to station them at Trelew, the northernmost airfield used in the war. This helped reduce the congestion at the Patagonian air bases operating the short-range aircraft. The C-130s and other civilian craft operated out of Comodoro Rivadavia, south of Trelew but still 860 kilometers [533 miles] from the islands.) Tierra del Fuego stretches east of Río Grande for more than 200 kilometers, yet apparently no consideration was given to using that landmass to shorten the flying distance to the islands. Leveling hills and blasting mountainsides in the rugged environment of Tierra del Fuego might have seemed prohibitively expensive, but it would have been a small price to pay for victory. Even an emergency strip for the recovery of damaged craft in Tierra del Fuego could have helped. An abundance of airstrips is one way to compensate for the lack of suitable airplanes, yet the air force took little initiative.

Everything indicates that local citizens were primarily responsible for improving the airfields in Patagonia and that where local support was scarce or nonexistent, the air force faltered. Thus, no construction work took place in Tierra del Fuego east of Río Grande. The failure to improve the existing runway at Port Stanley proved the most abysmal failure of all. Argentine workers had built this runway in 1972; originally, it was supposed to be long enough to handle all aircraft, but farsighted British officials had insisted on reducing the length to only 1,250 meters (4,101 feet) to preclude combat use. Thus, only civilian and cargo planes could take off comfortably from the runway. It was incumbent on the air force, and particularly on Lami Dozo, to put the whole weight of his office behind lengthening the runway as quickly as possible, no matter what the cost. In reality, the task received little attention, and tardy efforts failed to bring earthmoving and other construction equipment and materials to Port Stanley before the British imposed a naval blockade on the islands; no provision was made for construction workers either. Although military history traditionally places the emphasis on combat units, in the Falklands, victory depended more on construction crews than on ill-trained and demoralized conscripts. The construction work was so crucial that just adding access lanes or parking spaces meant immediately increasing the capacity of the runway to handle cargo planes from the mainland. From 2 April to 1 May, no more than six planes could use the airport simultaneously because of space and parking problems; planes parking off the concrete became mired.

After the war, Argentine officers came up with some lame excuses to try to justify the failure to undertake construction work in Port Stanley. For example, one objection to lengthening the runway was the expense of the project, but paying such a high price would have signaled to the world a real determination to hold on to the islands at all costs. Feverish and costly construction work was the only way Argentina could match Britain's determination to use its entire navy to recover the islands. Another excuse was that the islands had fewer days of good flying weather than Patagonia did. Having many planes stationed on the base might have been risky, but the capability to land and refuel combat planes in Port Stanley would have threatened the British recapture plan. Still another reason to ensure that construction workers, equipment, and material were readily available would be to repair any runway damage. A lengthened runway also would have forced British carriers to orbit farther away to escape deadly hits. Instead, by not even starting construction work on the runway, Argentina demonstrated both a scant desire to defend the islands and incomplete planning for the invasion.

The obvious British goal was to recover the islands, but in case of failure, the British were determined to at least force Argentina to pay a high price to keep them. The British respected overwhelming force and stubborn determination; consequently, the contest was as much about human will as about weapons systems. In the first days of April, the British government had decided to throw everything it had into the war; if later intelligence or initial combat engagements revealed that its assets were inadequate, Britain could always seek a diplomatic solution. In contrast, the junta had not made a similar commitment to throw everything Argentina had into the war, even though crash construction projects in Patagonia and on the islands likely would have forced the British to reconsider their invasion plans. The Argentine treasury minister received well-deserved criticism for not instituting a war economy immediately, but the military junta itself seemed unworried. To say that the junta lacked will does not go deep enough into the reason for the limited Argentine deployment and the partial mobilization of national resources.

The lack of will to commit all resources to the war had two fundamental causes. First, the military junta could not shake its belief that a peaceful solution to the crisis was the most likely outcome. This faith in a diplomatic solution rested largely on the false assumption that the United States would not allow a war to break out between two of its allies. The junta believed that the favors Argentina had done for the United States in the Cold War, particularly in the struggle against communism in Central America, had more than earned U.S. support for a negotiated solution. Whatever naive assumptions the junta had about U.S. support should have evaporated by 2 April, when the United States tried to persuade Argentina to accept the British terms. The U.S. position was obvious to anyone who bothered to read the newspapers, but the junta and the Foreign Ministry—trapped in a fantasy world—continued to believe that there was a peaceful way to keep the islands. The army leadership believed most strongly in the inevitability of a peaceful solution, and Galtieri shared this opinion. The belief that war was impossible reached down to the lowest army ranks. The post-invasion commander of the garrison in the islands, General Mario Benjamín Menéndez, who should have been feverishly preparing for the British counterattack, was instead reassuring his troops that this was all mock theater for a peaceful solution. The contrast between the army and the air force was particularly striking near the runway at Port Stanley. The air force personnel, using the few tools at their disposal, were furiously trying to erect barriers and construct air-raid shelters; they became the butt of jokes by the soldiers, who idled their time away. The jokes ceased when the bombs fell on 1 May; then the soldiers rushed to take cover in the shelters built by the air force. Even after 1 May, the junta stubbornly insisted on interpreting the hostilities as part of a diplomatic minuet rather than as the start of a real war. Not until the loss of Goose Green on 29 May did President Galtieri realize that the army was incapable of stopping the British and that Argentina needed to seek an evacuation through diplomacy. Unfortunately, his weak character kept him from imposing his viewpoint on the other junta members and on his subordinates, who still believed that Port Stanley was impregnable.

The second cause of the lack of will on the junta's part was the renewed fear of a two-front war. The extent of Chilean cooperation with the British is still unknown, but it was enough to cause the military junta to fear that Chile was preparing to strike Argentina in the back. Chile sent reinforcements to the south during the war and provided considerable intelligence to the British. Most notably, the initial intelligence on the whereabouts of the cruiser *General Belgrano* came from Chile. It is extremely doubtful that Chile ever intended to attack Argentina, but Chile served Britain more than adequately by bluffing Argentina with the threat of a two-front war. The junta should have considered the possibility of Chile cooperating with the British prior to 2 April, but even afterward, it could have used the Chilean threat as a pretext to justify a prompt evacuation of the islands, thus saving face and avoiding meaningless slaughter and destruction. Or Argentina could have called Chile's bluff and concentrated all its resources on the war with Britain. Instead, the military junta stupidly fell into the trap. The army kept its best units on the mainland to face a Chilean invasion; likewise, the navy sent only one of its superb battalions of marines to the islands. The Chilean bluff also influenced the navy's decision to withdraw its surface vessels from the war. The air force was the least affected by the Chilean bluff and proved to be the most willing to commit its air wings to the battle for the islands. Again, the air force had to try to compensate for the gaps left by the other two services when they refused to commit their best units to the defense of the islands.

In mid-April, for the first time in its history, the air force participated in trial runs against Argentine warships that were virtually identical to their British counterparts. Initial results of the mock attacks were encouraging. Accepted wisdom at the time called for no fewer than seven planes attacking simultaneously to sink a modern warship. The trial runs revealed that three and sometimes only two planes could bomb and sink a warship defended by even the most modern missile systems. Because the planes could not waste valuable fuel circling to establish large flight formations, this preliminary finding justified risking two or three airplanes in isolated attacks on British warships.

Unfortunately, time was too short for the air force to complete a full set of trial runs. Conclusive as the results were in terms of a purely warship versus airplane contest, the air force did not have the opportunity to determine how the presence of British Harrier planes would affect the defense of British warships. And in any case, not all the preliminary findings from the mock attacks were encouraging. The tests had revealed that to avoid the warships' radar, the attacking planes would have to fly very low, but at that altitude, fuel consumption increased and thus aggravated the already insufficient flying range.

Another disturbing reality did not need any tests to become blatantly obvious. In accordance with the decision of 1969, air force planes had never received equipment to navigate over the sea, so they could not plot their positions accurately. For unexplained reasons, some Skyhawks lacked any navigational equipment at all. In preparation for a war with Chile, pilots had familiarized themselves with landmarks to compensate for poor readings from their crude instruments, but those practices were useless out on the open sea. None of the planes had the most modern electronics, and most planes lacked even the simplest navigational radar. The solution with the Skyhawks was to have the planes with better navigational equipment guide the others; if, for some reason, any planes fell out of formation, the complications could be disastrous. For the Mirages and the Daggers, civilian Lear jets served as guides; this solution was not entirely satisfactory because, even with a considerable head start, the Lear jets could not keep up with the combat planes, even if the latter flew at their slowest speeds. Ingeniously, Lear jets often teamed up with the slow Skyhawks to give the British the impression of a larger attack force than was really the case. The Lear jets also ran at least 100 decoy missions that forced the Harriers to scramble to intercept them, thus diverting British warplanes from other targets or from the supply flights of the C-130 cargo planes to Port Stanley. The British eventually caught and shot down one Lear jet on a decoy mission.

With such fundamental navigational deficiencies, the radar at Port Stanley assumed an importance out of proportion to its traditional role. The airplanes are so glamorous that it is easy to forget that the Westinghouse AN/TPS-43 F radar at Port Stanley was the single most important piece of equipment for the air force and for the entire Argentine war effort. Successful operation of the radar did not guarantee victory, but its destruction would have paralyzed Argentine military operations almost immediately. Given that it was the Achilles' heel of the war effort, elementary precautions called for an additional radar set farther to the west. This second radar was also needed because the mountains blocked the radar signals from Port Stanley; the British fleet took advantage of this radar "shadow" to steam in undetected for the landings at San Carlos Bay on 21 May. Again, fear of Chile prevented the deployment of a second radar set, which, among other benefits, would have given ample warning of the approaching British fleet. Fortunately for the Argentineans, the repeated British attempts to destroy the main radar in Port Stanley all failed. Even the use of special antiradar missiles yielded the destruction of only one of the many smaller radars of the antiaircraft batteries. Thanks to the ingenuity and the resourcefulness of the radar personnel, this allowed Argentina to remain in the war.

Although the British certainly tried to knock out the radar, in a mistaken assessment of the situation, they concentrated their bombing efforts on attempting to destroy the runway. The British overestimated the Argentine capabilities and thus failed to realize that destroying the radar would have made the runway useless. For many reasons, the radar had acquired an inordinately crucial role. First, it compensated for the nautical deficiencies of the air force. The radar provided valuable information for the attacking planes—not just coordinates, but also warnings to abort missions because of nearby Harriers. The radar located many British ships and thus was the initial catalyst for bombing attacks. After 1 May, the radar safely guided every one of the huge C-130s that supplied the garrison. By carefully observing the behavior of the Harriers, radar personnel detected periods of relative inactivity (one was suspiciously near teatime), when it was safest for the C-130s to land in Port Stanley. The radar helped compensate not just for the lack of basic navigational equipment aboard the combat planes but also for the shortage of reconnaissance planes. Without sophisticated surveillance equipment or long-range reconnaissance planes, Argentina could not track the positions of British warships in any consistent way. One radar that was partly hemmed in by mountains could not provide full coverage of the vast sea around the islands, so to try to complete the picture, the air force improvised using the C-130s. The disadvantages of using this large, slow, bulky plane for combat reconnaissance can well be imagined. Yet, with the help of evasive techniques, the air force suffered the destruction of only one C-130 during the war. After considerable effort and extreme maneuvers, the British finally intercepted one of the planes, but even in that case, the British erroneously believed that the C-130 had been on a supply run, not a reconnaissance mission.

Lessons Learned from the Hostilities: 1 May to 14 June 1982

The British bombing of Port Stanley on 1 May marked the start of hostilities. That attack was the only time the antiaircraft defenders were caught by surprise. The greatest impact of the bombing raid of 1 May was the impression it made on the Argentine commander. For the rest of the war, General Menéndez held the unshakable conviction that the British intended to land at Port Stanley; consequently, he kept his troops deployed facing the sea and refused to fortify the heights to the west. The deadly Argentine barrage the next day persuaded the British planes to fly only at a high altitude and thus kept the release point for their bombs a long distance away from Port Stanley for the rest of the war. Not surprisingly, the bombs routinely missed their targets. To compensate for the ineffective bombardment, warships shelled the defenses every night. Although the British concentrated their shells on the tough marines, without any effect, the inability to stop the nightly bombardment hastened the demoralization of the army conscripts. Port Stanley itself remained safe from air attack. Even the Argentineans failed to appreciate the effectiveness of their radar-directed antiaircraft barrier, and foolishly, they dispersed their remaining ships and planes throughout the islands. Once those ships and planes were away from the defenses of Port Stanley, the British were able to destroy them in a piecemeal fashion.

The bombing of 1 May and the sinking of the cruiser *General Belgrano* at sea the next day set the course of the war for its duration. The belief that no combat would take place was not an army monopoly, so when a British submarine torpedoed the cruiser on 2 May, Admiral Anaya panicked. Fearing the loss of any more of his precious surface ships, Anaya used the excuse of the Chilean danger to keep the Argentine navy in port for the rest of the war. In contrast, the fleet commander had prepared a sensible plan to make skillful sorties with his warships against the British as soon as they attempted a landing on the islands, but Anaya's hasty and inexcusable retreat left full and undisputed control of the sea to the British. Furthermore, the self-imposed naval blockade rid the seas of all merchant ships after 1 May, thus limiting the garrison's supplies to whatever the air force C-130s could carry. The navy refused to even consider running the blockade with fast ships to bring supplies to the westernmost island, where the garrison was actually starving. For the service that had pressed the hardest for the occupation and defense of the islands to suddenly abandon the struggle was irresponsible in the extreme. The Argentine flight to the safety of home ports contrasted with the British determination to risk the entire Royal Navy, and in this test of will, London concluded that the operation to recover the islands could continue with every chance of success.

The abandonment of the islands by the surface vessels did not paralyze naval aviation. The Skyhawks aboard the aircraft carrier *25 de Mayo* and the five Super-Étendards redeployed to the airfields in Patagonia. To reach their British targets, naval airplanes needed aerial refueling from the air force's C-130s. On 4 May the Super-Étendards startled the world when their two Exocet air-to-ship missiles sank the British destroyer *Sheffield.* On 25 May two more Exocets crashed into and destroyed the *Atlantic Conveyor,* one of the vital British supply ships. The last of the five Exocets nearly hit the armored carrier *Invincible.* The Exocets had inaugurated a new era of naval warfare, but unfortunately, they were too few to turn the tide for Argentina. The navy had plenty of ship-to-ship Exocets, and after some ingenious rigging, it managed to fire one from Port Stanley on 12 June and hit the destroyer *Glamorgan,* one of the British warships shelling the Argentine marines nightly.

Having used up their air-to-ship Exocets, the Super-Étendards began to practice bombing runs, but the war ended before the pilots could apply their new skills. The Super-Étendards aimed to replace the Skyhawks lost by the navy during the 21 May attack at San Carlos. Although the navy Skyhawks were successful in their bombing of British ships, they suffered 50 percent losses during the escape. The destruction of the navy Skyhawks proved to have more momentous consequences for Argentina than did the loss of a larger number of air force Skyhawks. The navy planes were at least twice as valuable because their crews knew how to set the bombs properly for detonation against ships, whereas the air force crews lacked this expertise. By the end of May, the navy was, for all practical purposes, out of the war. The Fifth Marine Battalion remained as the primary naval contribution. The superb combat performance of this battalion, even in the face of heavy shelling, raises the question why the navy did not send another marine battalion to the islands in place of the utterly useless and even harmful army conscripts.

President Galtieri, as army commander, had already taken the precaution (later much criticized) of sending another army brigade to reinforce the garrison on the islands. The decision to reinforce was sound, in spite of the burden on the already stretched logistics; what was wrong was choosing to send perhaps the least-prepared conscripts from a subtropical region into the freezing islands. The ranking officers in the garrison, with few exceptions, were atrocious and set new levels of ineptness. The garrison commander, General Menéndez, was arguably the worst senior officer the Argentine army ever produced; his appointment revealed the rottenness deep inside the army promotion system. Among his many failures was not placing 155 mm cannon on the islands, first as a defense against British naval bombardment and later to offset the longer range of the British field artillery. The deplorable state of the army garrison somehow escaped the attention of the members of the military junta, who deluded themselves into believing that Port Stanley was impregnable. Operating under the false assumption that the garrison could continue to resist the British overland attacks, the air force believed that there was enough time to complete the feverish preparations to marshal new resources against the British fleet.

The air force was more than willing to fill the gap left by the departing navy, but Lami Dozo had to insist vigorously and firmly that the air force's sacrifice was justified only if the garrison was capable of sustaining a vigorous defense. The army's failure to attack the British bridgehead at San Carlos was not a good omen; nevertheless, rather than engaging in incriminations, the air force concentrated on overcoming its own limitations. The main constraint on the air force was not a lack of pilots but a lack of planes. Peru had sold Mirages to Argentina, and other warplanes were on their way; the air force expected to have all these planes operational by the middle of June. If the garrison could hold out until then, bad weather and a replenished air force were sure to halt the British offensive. If either the navy or the army supported the air force, Argentina could still pull off a victory. What was most remarkable was that the air force came very close to defeating the British without the help of the other two services.

Three Fatal Flaws

Lack of space prevents a description of even the most important of the 445 missions the air force flew during the war. Instead, this chapter concludes with three final observations on why flawed tactics deprived the Argentine air force of what could have been one of the most spectacular demonstrations of airpower in history. These are distinct from the previously mentioned errors in the initial phases of the air campaign.

The air force plan of operation of 7 April correctly assigned the highest priority in the bombing raids to British landing craft and troop transports. A judicious interpretation of the directive could include cargo ships, because from a combat perspective, supplies and equipment were just as valuable as the British troops themselves. The sinking of a troop transport with a huge loss of life was the fastest way to end the invasion, because electoral democracies have a very low tolerance for high casualties in regions far from a country's vital interests. Prime Minister Margaret Thatcher needed an easy and relatively bloodless conflict; high British casualties would wreck both the war and her political prospects. Argentine pilots have testified that they were ordered to avoid hitting troop transports; other pilots have admitted that they held back their bombs when cargo ships were the targets. In a possible contradiction of this version, one photograph shows bomb explosions straddling a troop transport; perhaps these were misses from bombs aimed at nearby warships. Whatever happened at the operational level in each mission, generally, air force pilots avoided attacking troop and cargo vessels. The pilots preferred to seek warships as the ideal targets, even though sinking many warships would not halt the landing. The British navy had decided to throw all its assets into this war and easily replaced the damaged or sunk warships. Instead, the loss of a few cargo and troop ships would have meant a halt of the invasion. These merchant ships were the Achilles' heel of the British effort, just as radar was for the Argentine side. The Argentine obsession with sinking warships rather than troop and cargo vessels not only deviated from the original air force plan but, more significantly, cost the country the war.

How the original plan to bomb landing craft and troop transports was transformed into an exclusive attack on warships is unclear. Sometimes pilots had only a few seconds to target whatever ship they saw. It is known that the Argentine pilots were unhappy about not being allowed to dogfight with the British Harriers. The air force command had concluded that the Mirage fighters operating at their extreme range did not have the time to engage the Harriers, which were armed with the latest air-to-air missile, the Sidewinder AIM-9L. The restriction on aerial combat irked the air force pilots, who perhaps expressed their frustration by attacking warships. The prohibition on aerial combat was well reasoned and not absolute. As the Web page of the Argentine air force noted in July 2003: "The problem was simple: If the Mirages descended [below their economical flight altitude], they consumed more fuel and could not return to their mainland bases. At that altitude, the pilots could stay ten minutes flying over the islands. Consequently, the conditions for aerial combat required that the Harriers climb, and that they do so in the initial minutes of the arrival of the Mirages over the islands" (translation by the author). Whatever the exact reasons may have been, a breakdown in the chain of command resulted in the needless sacrifice of planes and pilots to attack warships rather than the infinitely more valuable troop and cargo ships.

A second fatal flaw in the air force campaign was that a majority of the bombs failed to explode. Of the bombs that crashed into British ships, 60 percent did not detonate. A small margin of error for duds is inevitable in combat, but a figure of 60 percent reveals a major structural defect. In accordance with the directive of 1969, the air force had trained to drop bombs on land targets from a high altitude. The sudden change of mission in April forced the air force to bomb moving ships at sea. Worse, the planes had to fly below the radar near sea level and release their bombs at close range. At such low altitudes, the explosion risked destroying the plane as it barely managed to clear the ship's masts. Although the navy had developed delayed-action fuses, desperate experimentation by the air force failed to find a satisfactory solution in time. Given the rivalry and even hatred between the two services, the navy did not offer assistance, and the air force did not ask for it. As critics have correctly pointed out, rather than one unified war, Argentina was waging three wars, with each service conducting its own solitary campaign against the British. Cooperation and joint operations with another service took place only in the most extreme cases, such as the need for refueling. In Port Stanley, the British enemy forced the three services to cooperate, but even under fire, it was never adequate. On the mainland, with air force and navy planes generally stationed in separate bases, cooperation at the tactical level was virtually nonexistent. Had the 60 percent of duds exploded, the loss of warships and of at least a couple of transports would have forced the British fleet to abandon its invasion plans.

The third flaw involves a more traditional topic of debate about airpower: whether airplanes should attack in small numbers or in large formations. Obviously, changing such variables as the size and technology of each side affects the answer. In the specific case of the Argentine air force, the evidence strongly suggests that against British warships, massed formations were more effective than planes flying in pairs or small groups. When the Port Stanley radar reported that no Harriers were around, a large attack formation that confused the ships' defenses had the best chance of scoring significant hits. In contrast, planes coming in pairs at intervals gave the ships' gunners plenty of time to concentrate on their targets as they came on screen or into view. The least productive approach was a steady parade of planes; a series of planes coming singly or in pairs did not wear down the defenders and instead alerted the Harriers, which would arrive in time to either force the attackers to abort the mission or shoot down the Skyhawks before the latter dropped their bombs.

Many tactical events validate the superiority of large attack formations, but one mission on 8 June best exemplifies what might have been. Skyhawks had destroyed two troop and cargo vessels near a landing site that day, and the returning attack planes reported seeing one British frigate slightly out to sea covering the landing. Because lack of reconnaissance had left the air force short of targets for weeks, the high command authorized the release of Dagger fighter-bombers to make *one* pass at the lone warship. By the time the five Daggers reached the site, the British frigate *Plymouth* had changed position and, to the surprise of both the pilots and the British, the Daggers, flying at 900 kilometers (558 miles) per hour, narrowly avoided crashing into the ship as its crew frantically rushed to operate the antiaircraft defenses. Did that encounter constitute the one pass the planes had been authorized to make? The pilots knew that to keep losses down to one or two airplanes, warships normally had to be attacked by surprise, but with the element of surprise gone, past practice and strict orders dictated that the pilots abort the mission to avoid prohibitive losses. However, undaunted by the odds, the pilots executed a precise turn and pushed their machines to almost 1,000 kilometers (620 miles) per hour. The Daggers rushed through the antiaircraft fire, blasted the ship with their 30 mm cannon, and hit it with their bombs; amazingly, all the planes returned safely to base. The charge of the five Daggers showed that a mass formation could overwhelm the defenses of a warship, even one on alert. Obviously, the many navigational and operational problems, such as the shortage of refueling planes, made it difficult to mass large numbers of planes, but the air force should have striven for large formations as the best tactic. Perhaps quite fittingly, all four bombs that hit the target failed to explode, and although the damaged and burning ship was out of the war, the *Plymouth* had narrowly escaped being blown to bits.

Of the three flaws—the fixation on warships, the bombs' failure to explode, and the scant use of large attack formations—certainly at least one, the unexploded bombs, sufficed to turn the tide of the battle, and correcting any two of the three flaws could have halted the British invasion. Had the air force correctly used its available assets, airpower was perfectly capable of winning the war single-handedly. The seemingly impossible task of substituting for the navy and army was feasible. And if just one of the other two services had put up a decent effort, that, in combination with the air force, would have sufficed to halt the British invasion. The navy had originally planned to save its surface vessels for a sortie once the British landing began, and a determined navy raid certainly would have taken the pressure off the planes attacking San Carlos. Or if the garrison had vigorously attacked the bridgehead at San Carlos, the combination of land and air attacks would have doomed the British invasion. Instead, in the last week of the war, the air force desperately strove to compensate for the many deficiencies of the army, including inadequate artillery. To bolster the collapsing morale of the army garrison, the air force diverted resources to extremely risky ground support missions with its vulnerable Canberra jet bombers. Because the army did not know how to direct air support, the raids were largely ineffective, and in any case, the army garrison had lost its will to fight long before. Buenos Aires refused to accept the harsh reality of military collapse on the islands, and the surrender of the garrison on 14 June 1982 came as a complete shock. Up to the end, the air force believed that it still had time to get its new planes into the air and turn the tide of the war.

Suggestions for Further Research

The air campaign in the Falklands war has numerous unexplored areas. However, because of the huge obstacles blocking research, the field cannot be recommended to scholars. Unlike students of the periods up to World War II, no unpublished archival materials are available. Of the three armed services, the air force has been the most reluctant to grant scholars direct access to its personnel. It has partially compensated for this reticence by allowing the publication of narratives and interviews, but without the filter of an independent researcher.

For the near future, the most promising approach is to reconstruct the early history of the Argentine air force. No scholarly analysis of its history exists, so when the Argentine air force suddenly burst on the scene in 1982, there was no readily available perspective. Some archival materials are available to trace the birth of military aviation in Argentina and the development of a propeller-based air force. The transition to jet airplanes during the Juan Perón era (1946–1955) and the establishment of the air force as a service separate from the army await scholarly examination.

Once sources become available, military historians will be able to shed considerable light on the period from the fall of Perón in 1955 to the outbreak of the war in 1982. During those years, Argentina made the crucial decisions that shaped the military aviation that was available to face the British. A scholarly examination of the structure, functioning, and practices of the air force is sorely needed. The immediate prewar period from late 1981 to March 1982 was crucial. During those months, air force chief Basilio Lami Dozo knew that the military junta was planning to invade the islands but apparently did nothing to prepare the air force.

The whole issue of the leadership of Lami Dozo during the war remains a mystery. He has not satisfactorily explained his lack of action. Most notably, the plans and motivations behind the decisions made at the headquarters and air bases in Argentina remain unclear or largely unknown. The pilots and ground personnel have been willing to speak honestly about their successes and failures, but the administrators have kept silent about the key decisions. Many tactical and operational details also remain unclear; most glaringly, an exact operational inventory of the weapons, airplanes, and equipment of the air force is sorely needed.

Recommended Reading

The bibliography on the Falklands war is huge. The selection that follows represents my judgment about which books are particularly informative on the Argentine air force or the war itself. Although I have tried to list mainly English-language publications, most of the best books are in Spanish. Anyone wishing to gain a deeper understanding of the war needs a good command of that language.

B. H. Andrada, *Guerra aérea en las Malvinas* (Buenos Aires: Emecé Editores, 1983), draws on interviews with air force personnel to construct a narrative account of key engagements and episodes. This is an invaluable source.

Comisión Rattenbach, *Informe Rattenbach: Investigación confidencial sobre la conducción política y estratégico-militar de las fuerzas armadas argentinas en la guerra de Malvinas* (Buenos Aires: Ediciones Fin de Siglo, 2000), is a classic in the military history of Latin America. This book is the single most important publication in any language on the war. Although the commission did not delve deeply into its tactical side, this riveting account of gross incompetence at every level helps put the air force in its proper context.

Pablo Marcos Carballo, *Halcones sobre Malvinas* (Buenos Aires: Ediciones del Cruzamante, 1985), is the better of two good books by this author. This compilation of interviews with air force personnel involved in the war is an invaluable source.

Jeffrey Ethell and Alfred Price, *Air War South Atlantic* (New York: Macmillan, 1983), contains nice photographs, but the book is hurt by a lack of care. Much superior is the book by Jesús Romero Briasco and Salvador Mafe Huertas, *Malvinas, testigo de batallas* (Valencia, Spain: F. Domenech, 1984). This is the best book on the technical and operational side of the air war. The authors are Spaniards, and they enjoyed considerable assistance from their fellow NATO airmen in Britain and, more significantly, from Argentine airmen. This put them in a unique position to weigh the merits of the conflicting claims of each side. The book also has valuable sketches and abundant photographs.

Lawrence Freedman and Virginia Gamba-Stonehouse, *Signals of War: The Falklands Conflict of 1982* (Princeton, N.J.: Princeton University Press, 1991), is still the best overview of the war. It is very strong on the diplomatic and political aspects.

Max Hastings and Simon Jenkins, *The Battle for the Falklands* (New York: W. W. Norton, 1983), is the most reliable account from the British perspective. In spite of the authors' honest and conscientious efforts, however, some British propaganda slipped through.

Pío Matassi, *Probado en combate,* 2nd ed. (Buenos Aires: Editorial Halcón Cielo, 1994), contains well-illustrated accounts of combat missions. So far, this book is the best source on prewar conditions.

Rubén O. Moro, *The History of the South Atlantic Conflict: The War for the Malvinas* (New York: Praeger, 1989), is an English translation of the Spanish original and provides a good overall view of the war. The author is an Argentine air force officer and a veteran of the war.

In *The Fight for the Malvinas: The Argentine Forces in the Falklands War* (London: Viking, 1989), Martin Middlebrook makes a serious effort to present the activities of the Argentine military. Unfortunately, the air force refused to cooperate with him.

Robert L. Scheina, the author of *Latin America: A Naval History, 1810–1987* (Annapolis, Md.: Naval Institute Press, 1987), enjoyed the full cooperation of the Argentine navy. This highly recommended book is the best account in English on naval operations, including naval aviation.

Nigel West, *The Secret War for the Falklands: The SAS, MI6, and the War Whitehall Nearly Lost* (London: Little, Brown, 1997), uncovers some of the mysteries of the intelligence and espionage wars.

Videos

Videos can enhance one's understanding of the Falklands war. In contrast to the published works, most videos are in English; in fact, Spanish-language videos are few, not particularly informative, and virtually unobtainable in North America. A pro-British bias—if not outright British propaganda—tends to slip into these videos, which perhaps inevitably concentrate on the British role. Viewers have to make an extra effort to sift through the images to find the truth when watching videos made close to the events, such as *Falklands Task Force South* (1982). A more detached presentation is found in videos done some years after the war, such as *Battle for the Falklands* (1988). *Falklands: Soldier's Story* (2001) emerges as the best. All the videos make the indispensable contribution of depicting the barren terrain and the harsh weather in the islands.

CHAPTER NINE

From Disaster to Recovery

Russia's Air Forces in the Two World Wars

David R. Jones

Although Russia and the successor Soviet Union remained a Great Power, many, including scholars, have regarded it as a curious mix of backwardness and modernity. The bear has often been judged by its apparent military potential.

In the nineteenth century, military demands led to the creation of a small industrial base and railways, yet Russia was still overwhelmingly a peasant farmer state. After the revolution of 1905, the intelligentsia increasingly interested themselves in technology, especially aviation, where precedents had been set in the 1880s. An Aeronautical Commission was established in 1904; flying clubs followed, and after Blériot's 1909 cross-Channel flight, flying schools opened and French aircraft were imported. Official interest and the patronage of Grand Duke Aleksandr Mikhailovich of the navy propelled developments. Experience in the Balkan war of 1912–1913 also helped. Competitions were held to develop bombs, machine guns, airships, and aircraft. Igor Sikorsky designed and flew his four-engine Ilya Mourometz. In 1914 the three-year "large" armaments program included 1,000 aircraft. On the civilian side, enthusiasm peaked by 1913 because of costs and lack of suitable personnel. Nevertheless, upon mobilization in 1914, the tsarist forces could field 12 dirigibles and about 250 aircraft of some 16 types, adequate, it was thought, for the coming short war.

The School of War, 1914–1917

Unfortunately, the war was not short, and Russia's industrial backwardness soon undercut its aerial pioneers and negated the courage of its pilots. From the first, the War and Naval Ministries recognized the need to create a native aerial industry and intended that the Grand Program's massive orders would promote its development. Although the government frequently purchased foreign designs, the bulk of the new French Farman and other aircraft was to be built in Russia. Meanwhile, it sought to encourage domestic designers like Sikorsky with orders for both his and others' single- and multiengine machines. Partly as a result, a number of small airframe and aero-engine facilities sprang up, largely in St. Petersburg, Moscow, and Riga. These provided an industrial base for wartime expansion, but in August 1914, Russia's air services remained largely dependent on foreign aircraft and, in particular, on imported aero engines.

In early September 1914 the pioneer military flier P. N. Nesterov initiated aerial combat when he destroyed an Austrian aircraft (and himself) by the first *taran* (deliberate ramming). Although his exploit was celebrated, air-to-air encounters remained few and far between and played little part in the sharp decline in Russian airpower during the war's first four months. Rather, the decline resulted from the unexpected duration of the war, the high rate of attrition imposed by operational conditions, and shortages of replacement machines, spare parts, and engines due to the infant Russian aviation industry and an enemy blockade on further imports. In August 1914, for instance, Russia's air strength had included 224 French-designed machines (133 Nieuports and 91 Farmans); the seven major domestic producers were providing a mere 37 aircraft monthly, or some 400 for all of 1914, compared with Germany's reported 1,348. The latter, of course, had to support two major and (later) several minor fronts, but the Russians nonetheless found themselves at a considerable disadvantage. For instance, of the Grand Program's initial orders, only 242 of the expected 400 aircraft had left the assembly shops by October, including 7 of the 10 Sikorsky four-engine bombers. As for the large orders placed in allied France, by year's end, only 250 used aircraft and 268 engines had arrived.

Shortages in aircrew and maintenance personnel imposed further constraints on Russia's aerial operations. In 1914 the official pilot schools at Gatchina and Sevastopol claimed 130 graduates, for a total of some 300 since 1910–1911; the mobilization added those trained by the private aeroclubs. On this basis, the General Staff estimated that it had nearly two pilots for each available aircraft, which seemed sufficient for the expected short conflict. Despite an influx of volunteers (who were expected to provide their own aircraft and mechanics to service them), by the end of 1914 the air service complained of shortages in both fliers and observers. By early 1915, losses in both aircraft and aircrews threatened to doom the Imperial Air Fleet to impotence.

One bright spot was the performance of Sikorsky's four-engine reconnaissance bombers, built and promoted by former officer and industrialist M. V. Shidlovskii's Russo-Baltic Rail-Wagon Plant. Initially, aircraft and airships everywhere were employed mainly for artillery direction and reconnaissance, but in time, they increasingly carried out bombing missions as well. Although the Ilya Mourometz's debut in action in September 1914 was disappointing, Shidlovskii's lobbying prevented cancellation of the program and led to the formation of the Squadron of Flying Ships under his personal command. This squadron opened operations with raids against targets in the East Prussian rear in February 1915, and by October 1917, it had made 442 raids, unloaded some 2,000 bombs, and taken more than 7,000 aerial photographs. In addition, gunners on these aircraft downed some forty enemy fighters, while the squadron lost only three of its own planes in combat. The squadron's active strength usually averaged twenty-five machines at any given moment. Shidlovskii's plant provided a total of seventy-six such bombers and the necessary servicing and aircrew training.

The use of aircraft as well as airships for both reconnaissance and bombing missions naturally invited countermeasures, and pilots of all nations employed a variety of primitive means to bring down their opposites. Although serious air-to-air combat became possible only with the invention of a synchronizing mechanism that permitted firing through the propeller arc, "pushers" had mounted machine guns in a forward-observer's position. Such aircraft could also attack enemy airships, and the army again took steps to select the most suitable weapons for both Sikorsky's bombers and its single-engine machines. As a result, orders went out for British Lewis and American Colt-Browning machine-guns; pending their arrival, Russian pilots continued to use the lighter Madsens (ordered on the eve of the war) and experimented with other models. In addition, by late 1915–early 1916, fighters were being fitted with synchronizing mechanisms—of both domestic and foreign design—that allowed their pilots to aim the plane and fire directly at an aerial target.

Meanwhile, in early 1915, German air raids against Warsaw led to the formation of a detachment, armed with Farman pushers and dedicated to the city's defense, that was hailed as the first Russian "fighter" unit. But this branch achieved real recognition only in June–July 1916, when the active army's director of aviation and aeronautics, reviewing the successes of the fighter "circuses" on the Western Front, won approval for the creation of similar units in the east. Consequently, that August saw the formation of the First Fighter Group to serve the Eleventh Army, and in September, a second group was formed on the Southwestern Front. The latter was led by the ace E. N. Krutin. He had visited France to study Allied efforts and, on his return, had composed a pamphlet on the appropriate tactical use of such units. When this process was cut short in May 1917 by the Revolution, Russia had four fighter groups with a total of twelve squadrons and its own celebrated aces.

By that date, Russia's military air fleet had clearly recovered from the near collapse of 1915 and had achieved a degree of autonomy within the military-administrative and command structure. In 1916 overall control was delegated to Grand Duke Aleksandr Mikhailovich, who took charge of a new Main Directorate of the Military Aerial Fleet within the War Ministry. Frontline units operated under a director of aviation and aeronautics for the active army at *Stavka* (headquarters), whose position was transformed into head of a full-fledged Directorate of the

Field Inspectorate of the Air Fleet in early December 1916. Although theoretical debates still raged over the proper role for military aviation, these administrative changes signaled that it was at last becoming an "armed service" in its own right.

The air fleet's recovery was only one aspect of the revitalized war effort and the industrial mobilization initiated by the imperial government's creation of the Special Conferences for State Defense in the spring and summer of 1915. With the influx of new orders, Russia's seven aviation plants expanded to eigh teen; eleven built airframes, five built engines, and two produced propellers. Many were still small, but production reportedly increased to 205 planes a month in 1915 and to 325 by February 1917, for a total of 2,200. Most were still French Farman, Spad, Morane, and Nieuport designs built under license. But these figures also include Sikorsky RBVZ-17 and -20 fighters, as well as his bombers, Lebedev 7–12 fighters, and D. P. Grigorovich's M-9 naval flying boats, which proved especially useful.

All in all, Russia's air services received 3,100 new aircraft during the war, of which 900 were imported from France and Britain. Russia's domestic industry also developed and produced sufficient quantities of V. V. Oranovskii's high-explosive (eight-caliber) and fragmentation (five-caliber) bombs and, after 1916, A. Yakovlev's incendiaries for use by the tsar's airmen. Despite competition with ground units, machine guns were also available for arming the new aircraft, and Russian designers provided their fliers with bomb and gun sights that were as effective as those of their opponents. Only with regard to aero engines, which were considerably more complex to produce than airframes, did Russian manufacturers fail the service.

The dimensions of this failure are indicated by a report that 1,769 aircraft left the assembly lines in 1916, but only a paltry 666 power plants. Since monthly engine production had reached only 110 to 150 by early 1917, this shortage presented a real bottleneck that bedeviled Russia's aerial planners and forced them to power even domestically built airframes with foreign engines, reportedly importing 3,600 of them by October 1917. Nonetheless, by that February, some squadrons reported that they had only two engines for every five or six airframes. At the time, the Northern Front had only 60 fighters fit for service out of a formal establishment of 118, the Western Front complained of a 59 percent shortage of fighters, and other fronts were similarly hampered. This situation explains much of the celebrated opposition of Mikhailovich and his staff to expanding the Squadron of Flying Ships. As his adjutant for fighter aviation later explained to me, the air fleet lost four fighters every time a Sikorsky bomber entered service.

As suggested earlier, the other constraint on imperial Russia's wartime aviation was the shortage of technically trained manpower suitable for utilization as aircrew and maintenance personnel at the front and as a labor force in an expanding aviation industry. A September 1916 report claimed that the four official schools had trained a total of 446 fliers, plus those in the navy and graduates of unofficial programs. By 1917, the air fleet had some 500 fliers in active units. Meanwhile, despite attempts to recruit artillery officers, of the 357 aerial observers serving in mid-1916, only 132 were qualified artillerymen. Given the aircrews' accelerated training and the poor maintenance of frontline aircraft, losses remained high (25 to 30 percent of commissioned ranks), and reserves of qualified replacements were diminishing.

Russia's naval aviation had developed so that, by 1917, the Baltic and Black Sea Fleets each officially had two brigades composed of three divisions (18 aircraft per division) apiece. The Black Sea Fleet had over 152 combat-ready aircraft, but the Baltic had only 88. All in all, by early 1917, the navy reported a total of 493 aircraft (seaplanes included), but it too was hampered by engine shortages. Even so, by that date, the naval command was using aircraft for photoreconnaissance of enemy shore installations, searches for enemy surface ships and submarines, and offensive and defensive combat operations, and plans were under way for the design and development of bombers and torpedo aircraft for strikes on enemy warships. Some 3,000 officers and men (300 being pilots) served during World War I.

Taken in isolation, the story of Russia's wartime air services may seem uninspiring. Although major problems remained, and the Imperial Air Fleet never won full "air control," it was hardly crippled by the empire's legendary "backwardness" or allegedly corrupt and obscurantist leadership. From 1916 it held its own against its Austro-German opponents and successfully carried out most of its assigned missions. In 1916 Brusilov's shock troops trained on accurate models of the positions they would later overwhelm. This process was repeated before the disastrous Kerensky Offensive of June 1917, but by August, most units had only 50 percent of their initial equipment levels. Thereafter, the air services, like the rest of Russia's armed forces, rapidly disintegrated thanks to the corrosive influences of the Revolution. As aircraft and engine production ground to a halt, the air fleet's maintenance and supply nets fell apart, and its frontline units, riven by political divisions, lost their combat capability.

The Soviet Renaissance

The air fleet's wartime progress was virtually wiped out by the Revolution of 1917 and the subsequent civil war (1918–1920). Despite efforts to the contrary, the new Soviet regime managed to preserve only some 37 percent of its predecessor's personnel and materiel. This served as a basis for the Workers' and Peasants' Aerial Fleet, formed officially with the outbreak of full-scale civil war in late May 1918. By November 1919, it had a steady contingent of 350 machines in 62 squadrons. Trained personnel were in equally short supply. Despite the school's reopening in April 1918, the Red Aerial Fleet received a mere 155 pilots and 75 observers; though they were generally enthusiastic and dedicated, they lacked competence. Consequently, of the 422 aircraft lost between April 1918 and 1 April 1920, 365 were destroyed in accidents, and the scale of Soviet air operations was limited at best.

The new authorities also made strenuous efforts to reorganize the aviation industry, and in late 1919, these culminated in the creation of its Main Administration, or Glavkoavia. It produced only 255 machines and 79 engines during 1918; although matters subsequently improved, Soviet plants turned out only 558 new aircraft and 81 engines during

1918–1920. Equally worrying for future development was the emigration of Sikorsky, Severskii, and most of the aviation engineers formerly centered in Kiev and St. Petersburg. As a result, the Soviet government could rely only on N. Y. Zhukovskii's Moscow circle, the members of which had found escape more difficult.

Yet the new air services were fortunate, in that many of the victorious Bolsheviks, not unlike the "futurists" who supported Mussolini's Fascists, saw aviation as an essential symbol of the modernization they wished for Russia. Indeed, V. I. Lenin encouraged Zhukovskii and his fellows to create Moscow's Central State Aero-Hydrodynamic Institute, and the military to resurrect the Red Aerial Fleet.

The Soviet regime was determined to eliminate the weaknesses of the imperial air services. In late 1921–early 1922, this led planners to allocate resources to research and development and to expand both technical education in general and the production base of the air industry in particular. This was reflected in Joseph Stalin's first Five-Year Plan in 1928. Despite the startup of production of the R-1 reconnaissance biplane in 1923, during 1922–1924, the air fleet's strength was sustained largely by importing some 300 machines from Germany, Holland, and Italy. Close cooperation established with Weimar Germany's military after 1922 stimulated Russia's domestic industry, but production remained negligible. Indeed, during the period 1918–1928, Soviet facilities provided the state with a mere 2,847 planes and 1,930 engines.

This effort paralleled more successful programs to train the engineers and other personnel required by the new Red Air Force, through both professional educational institutions and a network of "voluntary" defense organizations, including aeroclubs. Dispersed by the Revolution, these revived slowly in the early 1920s. Gliding, thanks to its relative lack of cost, was the first aerial sport to win a wide following, and in January 1923 the USSR held its first competition in glider design and construction, as part of the structure of the Society of Friends of the Air Fleet.

This was only one of several such "voluntary" societies promoted by the Soviet state, the Communist Party, and the Young Communist League (Komsomol). Like the tsarist regime of 1910, they sought public and private donations to fund the creation of a pool from which the Red Army could draw the future pilots and technicians needed for a vastly expanded air fleet. To reduce the competition for scarce monies, in May 1925 the Society of Friends of the Air Fleet and its 2 million members merged with the Society of Friends of Chemical Defense and the Chemical Industry to form the 3 million–strong Society of the Aviation–Chemical Industry of the USSR (Aviakhim). Then, in January 1927, this group combined with the newly formed (1926) Society for Promoting Defense to create the Society of Friends of Defense and the Aviation–Chemical Industry of the USSR (Osoaviakhim).

This reorganization ended the societies' competition for members and funds, centralized their bureaucratic structures and the direction of organizational programs, channeled local interests and efforts more effectively, and brought under control the organizational chaos that had weakened their input to the defense effort. In addition, a major new element was introduced into Osoaviakhim's activities by the emphasis now placed on military training. Until 1927, Aviakhim's activities had concentrated largely on political agitation and programs of mass aerial propaganda, thanks to which the Soviet Union seemed to be one of the most air-minded nations of the day. Even so, its leaders admitted that although millions had enrolled by 1929, only 20 percent could be called active. As of 1932, Osoaviakhim still had only two real pilot schools.

Despite the new military emphasis and the educational and political benefits brought by these activities, only limited practical training could be provided with aircraft as a result of the Five-Year Plans introduced in 1928–1929. Since one major goal of this forced industrialization was the strengthening of the USSR's defense potential in general, and of the air forces in particular, all aspects of the aviation industry were reorganized and expanded. Backed by a revamped system of new design bureaus, a well-funded and expanding web of factories now struggled under the Main Administration of the Aviation Industry to put new generations of modern Russian-designed aircraft and engines into serial production. Despite a number of mistakes and horrors, the program managed to increase annual production to 2,509 planes by

1932 and, despite the negative impact of Stalin's purges, 10,342 by 1939. All in all, during 1929–1939, Soviet plants turned out a total of 45,499 aircraft, an impressive accomplishment compared with 1918–1928 production figures. Equally striking, by the mid-1930s, these included the modern I-15 and I-16 fighters, as well as the SB bombers. These and other models were armed with modern machine guns and bombs, and all performed well in the early stages of the Spanish Civil War (1936–1939) and against the Japanese in Manchuria in 1939.

The expansion of a pool of technically educated personnel to staff the air forces and Russia's new research, design, and production capabilities was equally significant. Official recognition came on 25 January 1931 when the Ninth Komsomol Congress enthusiastically adopted the proposal of People's Commissar of Defense K. E. Voroshilov and accepted formal patronage of the air forces (aerial fleet). Under the slogan "*Komsomolets—na samolet!*" (Komsomol member—to an airplane), the congress pledged to provide the air services with 150,000 trained pilots from the service schools and aeroclubs. Close cooperation immediately developed between the Komsomol and Osoaviakhim, and the number of aeroclubs expanded to twenty. This impetus was reinforced by the joint party–government decree "On Osoaviakhim" of 8 August 1935, which demanded that the society expand all aspects of military training. As a result of the so-called Komsomol pass or ticket *(putevka),* thousands of young men and women joined Osoaviakhim's growing network of aviation schools, air clubs, and circles, and 30,000 Communists and Komsomoltsy entered the air fleet's regular schools from 1931 to 1936.

Osoaviakhim's practical work grew accordingly. Thus, the pair of real flying schools and the single gliding school of 1931–1932 reportedly multiplied to 8 and 115, respectively, by 1934. The structure supporting these last also became much more dense: the number of glider stations went from 32 in 1931 to 170 in 1934, and these in turn were backed by some 800 glider circles. As a result, the number of sportsmen–glider pilots reportedly rose dramatically during the same period—from 500 to perhaps 17,000—and Soviet citizens captured thirteen of the eighteen international gliding records registered by the Féderation

Aeronautique International (FAI) in 1939. As for aircraft pilots, by mid-decade, student numbers had increased sevenfold from the 900 of 1932; this reflected the increased training aircraft available. During 1931–1936, the society's aircraft park grew by nineteen times and, with semi-official support, an aeroclub appeared in almost every major industrial center, as well is in many smaller cities.

Equally important, work began in 1933 at Moscow's Tushino Field on facilities for the Central Aeroclub of the USSR. Opened in 1935, it quickly federated with the FAI. In 1936 the Soviet government assigned this club responsibility for registering aviation records, and it soon became the main methodological center, as well as one of the best educational institutions, in the Soviet Union. It established its own research institute, which held its first graduation in 1934; it later became famous as the Central Institute for Aircraft Motors. In December 1938 the club was renamed the Chkalov Central Aeroclub, after the famous long-distance flier and Hero of the Soviet Union of the mid-1930s. Its graduates included noted pilots and aces, as well as the aircraft designers S. V. Il'iushin and A. S. Yakovlev.

After 1931, Osoaviakhim also promoted a range of other aviation-related sports, including parachute training. In 1933 it founded its Higher Parachute School, and in 1938 Osoaviakhim reported that its members had made 3 million jumps that year alone. Meanwhile, the sport of aircraft modeling spread even more rapidly, and on 31 August 1938 Osoaviakhim opened its Central Aircraft Modeling School at Veshniaka near Moscow. All in all, by 1 January 1941, Osoaviakhim claimed 2.6 million members enrolled in a network of some 800 training installations devoted to a range of defense-related activities. These included automobile, radio, cavalry, and naval clubs; schools of air defense and other military training centers; and some 180 aeroclubs and 46 glider schools.

From 1930 to 1941, these clubs reportedly trained 121,000 aircraft pilots and 27,000 glider pilots, which clearly benefited both the civilian and the military services. In the latter, for example, the number of air force officers rose from 13,000 in 1937 to 60,000 in 1940, and by 1941, that service maintained seventy-eight flying schools, eighteen engineering schools, and three academies. Such figures seemingly justified the Soviet claims of Russia's new "air-mindedness" and demonstrate the success of the USSR's state-sponsored, paramilitary "sports" program.

Apart from the combat feats of some of "Stalin's Falcons" over Spain and China, other fliers won fame by setting international records, especially those involved in a nonstop flight from Moscow to New York and in the landing of a Soviet plane at the North Pole. By 1939, the Red Air Force was recognized as the world's largest and, given the reserve pool created by the paramilitary programs just described, apparently faced an even brighter future.

The Disaster of 22 June 1941

Unfortunately, deep flaws existed beneath this superficial strength. These became obvious on Sunday, 22 June 1941, when Hitler's Luftwaffe inflicted one of the most devastating defeats in aviation history on the Red Air Force. By that date, the latter's ranks had swelled to 400,000 personnel and 10,000 to 15,000 aircraft. Not all, of course, were suitable for combat, and only 7,500 were deployed in the Soviets' western theater. There they faced the 2,800 (out of 4,300 total) combat aircraft of the four German *Luftflotten* (air fleets) assigned to Operation Barbarossa. Consequently, the Luftwaffe sought command of the air by first destroying its enemy's numerical superiority, and on 22 June, some 1,000 bombers launched repeated strikes against the sixty-six fields in Russia's three western border districts.

Thanks to initial surprise, the results were nothing short of spectacular. Figures on the actual losses vary, but by noon, some sources claim that 890 Soviet planes had been destroyed (668 on the ground). Indeed, the Germans claimed that on the first day they destroyed a total of 1,811 enemy planes, 1,489 of which never became airborne, at a cost of only 35 of their own machines. In the 1970s a Soviet "official history" admitted the loss of some 1,200 aircraft, 800 of which were destroyed on the ground. Whatever the case, the initial German assaults crippled the Soviet frontline air forces by 47 percent (738 of 1,560 aircraft) and their commander (who committed suicide). The aviation arm of the Northwestern Front was left virtually helpless, and even that of the less hard-pressed Southwestern Front lost 277 combat machines.

The same strikes also utterly disrupted the Red Air Force's logistical and support networks and created chaos within their command-and-control agencies, a process that was furthered by Stalin's execution of some senior leaders as scapegoats. By day three, the Luftwaffe was therefore free to support the rapidly advancing German Panzers. These then seized the Soviets' forward airfields, along with all the disabled aircraft awaiting repair and the massive prewar stockpiles of bombs and other munitions. Those Soviet pilots who fought back had little if any tactical, let alone operational, control, and they frequently found themselves forced into suicidal strikes. As a result, Russian losses continued to mount. By 30 June, the Western Front had lost 1,163 planes, or 74 percent of its initial strength; German headquarters claimed the destruction of a total of 4,017 Soviet machines, at a cost of 150 of its own. This total reached 6,857 by 12 July, and three days earlier, the Soviets themselves admitted the destruction of 3,985 aircraft. And despite the new *Stavka*'s efforts to conserve its dwindling resources, post-Soviet Russian figures verify that from 22 June to 31 December 1941, the Red Air Force lost a staggering 10,300 combat aircraft out of a total of 17,900.

Stalin and the Debacle of 1941

Given the efforts devoted to developing military aviation after 1929, the causes of this disaster deserve attention. Overall, they can be divided into the "immediate" and, more fundamentally, the "structural." In both categories, Stalin, the unquestioned architect of Soviet policies at all levels, emerges as the culprit. Although other factors played a major role, the extent of the disasters of 1941 in general, and of the catastrophe that engulfed the Red Air Force in particular, is a prime example of the malevolent impact a leader can exert on his nation's history.

For close observers, both Soviet and foreign, these disasters were less than startling. Soviet pilots had continued to score successes in the Far East during 1938–1939, but the arrival in Spain of new German aircraft (e.g., the Messerschmitt Bf 109) revealed that the Russians' air capabilities were declining. Many in the Soviet air establishment downplayed events there, but they could not ignore their services' abysmal performance against little Finland in the Winter War of 1939–1940. After much debate and some confusion, Stalin and his commander approved a major program of reorganization and rearmament on 25 February 1941, which was well under way at the time of the German assault and thus an immediate cause of the June disaster.

Drawing on the perceived lessons of Spain and conflicts elsewhere, this program restructured the air forces to stress the tactical role of frontal aviation (as opposed to long-range strategic missions). Among other things, it also expanded the composition of air regiments, reorganized the rear services, ordered the creation of 106 new air regiments, and, accordingly, called for the expanded training of aircrews by both the air forces and Osoaviakhim. This reorganization was accompanied by a reequipment program aimed at replacing the old I-15 and I-16 fighters and SB-2 bombers with the new generation of Mig-1, Yak-1, and LaGG-3 fighters, Pe-2 and Il-2 ground attackers, and DB-3F/Il-4 medium bombers. Serial production of these new planes began in 1939–1940, and in 1940 A. I. Shakhurin took over an expanding Ministry of the Aviation Industry to speed up the process. By 1941, he controlled fifteen plants producing fighters and light machines and nine devoted to bombing and ground attack aircraft.

Such a total overhaul obviously required time, and on 22 June, it was still under way. Only 19 of the planned 106 new air regiments had been formed, problems in the quality of training persisted, and, in the wake of the Finnish war, morale was uncertain. One report maintains that, on the eve of war, the air services faced a 32 percent shortage in aircrew and technical personnel. Since there was general agreement that the program begun in February 1941 could be completed only by mid- to late 1942, the air forces' higher commanders were naturally uncertain about how to fulfill a newly defined set of missions and, given recent demotions and appointments, about their own and their service's future prospects. Given that similar problems existed in other branches of the armed forces as well, Stalin cannot be faulted for hoping to postpone a conflict with Nazi Germany at least until 1942, nor can he be blamed directly for many of the practical and technical problems involved in the ongoing reorganization.

This said, other policies of the Soviet leader clearly complicated and retarded the restructuring process and otherwise helped maximize the Luftwaffe's initial successes. First and foremost are his acquisition of eastern Poland and the Baltic republics after the August 1939 pact with Hitler and the resulting forward deployments required to defend this border glacis. Whatever his strategic motives, this imposed a major additional strain on the resources of an air force that already desperately required reorganization and reequipment. Thereafter, for example, two-thirds (or some 200) of the airfields built or renovated were in the newly expanded western districts to support the aerial concentrations required to win air control in a future conflict.

Because this work remained unfinished in June 1941, the newly enlarged units were often deployed on crowded primary airstrips or unfinished airfields, which complicated efforts at dispersal or camouflage. This, along with the lack of effective early warning, left them vulnerable to surprise strikes. The danger was heightened further by Stalin's unwillingness to face the reality of the coming invasion. Space precludes a discussion of the dictator's motives in ignoring numerous warnings, his banning of attempts to intercept the steadily mounting number of German reconnaissance overflights, or his unwillingness to order timely defensive measures. Suffice it to say, he finally issued an order for the dispersal and camouflaging of aircraft on 19 June, much too late to diminish the coming catastrophe. His desire to avoid a conflict was understandable, but his refusal to recognize the looming threat did much to facilitate the Luftwaffe's victory.

Stalin was directly responsible for the disruptions plaguing the air force command in the years preceding the German assault, and he was indirectly responsible for the theoretical confusion and uncertainty that frequently hindered planning for a future conflict and inhibited a more timely reorganization and reequipment of the air services. His motives for the purges that struck the armed forces after May–June 1937 may still be debatable, but their impact is not. In the first wave, the military as a whole lost 15,000 to 30,000 of its 75,000 officers. Within this context, the air forces were especially hard hit: 4,724 (or 36 percent) of its 13,000 officers were arrested in 1937, and a total of 5,616 by January 1940. These included 75 percent of the most senior and experienced commanders, the air force chiefs included, along with the leading theorists on aerial warfare and proponents of strategic airpower (A. N. Lapchinskii, A. C. Alazin, and A. K. Mednis). Air services commander Y. I. Alksnis and his deputy disappeared in 1937, his uninspiring successor A. D. Loktionov in September 1939, the Spanish veteran Y. I. Smushkevich in August 1940, and P. V. Rychagov eight months later in April 1941. All subsequently died in prison or were executed.

Rychagov in particular is symbolic of the inexperience and declining quality of the senior command on the eve of war. Although he took charge when the Red Air Force was assessing its failures over Finland, he was only twenty-nine years old and had been a mere lieutenant only three years earlier. Although it is true that many of the arrested military officers were soon rehabilitated and returned to the ranks, in the air forces, only 16 percent (892) of those purged by 1940 were so fortunate. It is small wonder, then, that morale was low. Stalin's clear preference for political reliability (or conformity) over professionalism led to questionable combat capabilities within an officer corps comprised at all levels of purge survivors, hastily promoted and inexperienced youngsters, and recently recruited newcomers. For this, too, Stalin must take full blame.

His purges were equally disruptive for the aircraft industry. Although the need for new aircraft with heavier weapons had become evident in Spain during 1937, neither the aircraft nor the armaments industries were spared by Stalin's secret police. After an initial wave of arrests, the hunt for "saboteurs" and other "enemies of the people" spread throughout the research institutes, design bureaus, and production facilities. At the Aero-Hydrodynamic Institute, the purge moved from the top of the administration down to the level of clerks and even briefly involved the director, S. V. Chaplygin. Repeated in every institute, bureau, and factory, this process brought the dismissal, arrest, and often execution of thousands. Among those killed were the recoilless cannon designer L. V. Kurchevskii, the rocket pioneers I. T. Kleimenov and G. E. Langemak, and the talented aircraft designer K. A. Kalinin. Others were more fortunate and, like the designers A. N. Tupolev and V. M. Petliakov, continued to work in special prison bureaus under the supervision of the People's Commissariat of Internal Affairs (NKVD).

Meanwhile, the surviving designers and industrialists met in early 1938 to initiate a program that would produce a new generation of aircraft dedicated to close- and medium-range air support of ground forces, along with the heavier armament required for these missions. Despite the uncertain conditions, by the end of 1940, the bureaus had designed, built, and flight-tested 115 original and 83 modified aircraft. Of these, 20 were ordered into serial production within an expanding aviation industry, which itself was reorganized in January 1940. Other designers provided the new weapons (the UB heavy 12.7 mm machine gun and 20 mm ShVAK cannon) and improved bombs with which to arm these new aircraft. Yet here, too, the purges had their impact. Since economic planners who set production "norms" often overlooked the time needed to resolve the inescapable problems encountered in retooling assembly lines or to correct defects in hastily approved prototypes, many factory managers dragged their heels when it came to making the necessary conversions. For them, it was safer to produce older models and reach their quotas than to risk charges of "wrecking" by pausing to convert their plants to produce more modern aircraft.

Consequently, from January 1939 to 22 June 1941, a total of 13,852 older fighters and bombers left the assembly lines, compared with 2,839 of the new fighters and ground attack planes and 2,065 DB-3/Il-4 bombers. Still more informative are the figures for 1940: 7,267 old fighters and bombers versus 186 new fighters and ground attack machines. The latter's production numbers then rose rapidly to 2,653, as opposed to only 657 older types, during January–22 June 1941. Although a reported 1,540 modern warplanes reached the western districts, time was required for flight conversion training. Again, thanks to the purges, this went slowly due to fears that accidents would bring charges of "sabotage" or "wrecking." Eighty percent of the Mig-3 fighter and 72 percent of the Pe-2 light bomber crews had completed retraining by 1 May 1941, but only 32 percent of those assigned to the LaGG-3 fighters had done so. It is thus no surprise that most of the new machines assigned to the frontier districts were lost in the subsequent catastrophe. By November 1941, both German analysts and most outside observers had concluded that the Red Air Force was incapable of seriously opposing, let alone defeating, the victorious Luftwaffe.

The Soviet Recovery, Summer 1941–Winter 1942

One year later, this picture had changed abruptly. By August 1942, the German Sixth Army had reached the outskirts of Stalingrad on the Volga, and by November, it was battling to seize the city proper. The advance had created a bulging salient, the flanks of which offered a tempting target for planners at Stalin's *Stavka.* The result was Operation Uranus, under which troops of the Soviet Stalingrad and Southwestern Fronts struck on 19 November 1942. Thanks to the element of surprise, by the twenty-third, they had trapped some 255,000 Germans and 12,000 allied Hungarians and Romanians within what became known as "Fortress Stalingrad." Refusing to permit a breakout, Hitler instead ordered his Luftwaffe to supply the beleaguered Sixth Army by an airlift. The Red Air Force responded by establishing a "ramified aerial blockade." Two months of vicious air battles followed until the garrison's starving survivors finally surrendered on 2 February 1943. These battles, as Luftwaffe general D. W. Schwabedissen later admitted, "proved clearly that Russian aviation matched that of the Germans who had lost their earlier air superiority. . . . After the beginning of 1943 the Luftwaffe was mostly on the defensive."

The Turning Point

In reality, signs of a Soviet revival were apparent much earlier. Although Soviet losses in the summer of 1941 had been stupendous, those of the

Luftwaffe were far from negligible. By early August, it had lost 1,023 of its initial 2,800 aircraft, with another 657 undergoing repairs. Nonetheless, by 30 September 1941, the Russians could deploy only 545 largely obsolete machines to oppose the 1,000-plus Luftwaffe aircraft supporting the drive on Moscow. Determined to reduce this advantage, in October and November *Stavka* ordered antiair strikes against German airfields. Although these apparently achieved significant results, more important was the arrival of four aviation divisions (more than 1,000 aircraft) from the Far East, where the Japanese threat had waned. Consequently, in mid-November the Russians massed 1,138 machines (738 active) to oppose the Luftwaffe's 670. But even though the Red Air Force had gained numerical superiority, the experienced Luftwaffe continued to hold its own, and then some, until autumn 1942.

As of 19 November, the Germans remained confident that their Fourth Air Fleet had the upper hand on the Stalingrad Front. At that time it fielded some 350 fighters, 550 bombers, and 100 reconnaissance planes and had five air transport groups of Ju 52s and one with He 111s. Against them the Soviets had units with a reported inventory of 1,916 combat aircraft of all types, 1,350 of which were fit for service on 20 November: 509 fighters, 420 ground attack aircraft, 90 bombers, 308 night bombers, and 23 reconnaissance planes. The Luftwaffe suspected that it was outnumbered but still counted on the perceived qualitative edge of its men and machines to guarantee continued air superiority. Indeed, in October 1942 Hitler declared that Russia was weaker than it had been in the desperate winter of 1941–1942.

These assumptions began dissipating on 19 November. That day found the Luftwaffe completely grounded by sleet, fog, and snow. By 23 November, it had managed only 361 sorties, which meant that the weather had virtually deprived the German ground forces of air support. Nature played no favorites, but the Red Air Force did somewhat better. It managed only 44 missions on the nineteenth and none on the twentieth, but it had flown a total of 369 by the twenty-third and 970 by midnight on the twenty-fourth. After this, the pace quickened; by 30 November, Soviet commanders claimed a total of 3,769 sorties (raised by some postwar writers to 5,760), compared with some 2,000 by the Germans. Meanwhile, the Germans were organizing their airlift, and this gathered momentum during December 1942, but so did the tempo of Red air operations. In that month, the Eighth and Sixteenth Air Armies alone flew a total of 10,459 missions, including 1,838 strikes against enemy troops, 1,757 escort missions, and 4,147 sorties aimed at destroying enemy aircraft (airlift transports included); these last involved 2,856 assaults on German airfields, 273 offensive patrols, 468 dedicated intercepts, and 550 "free hunts" by Soviet aces. If nothing else, this diversity is clear evidence of the growing sophistication of the Soviets' resurrected air services.

Given the disasters of 1941 and the earlier defeats in 1942, this recovery at first seems nothing short of miraculous. In fact, it resulted from a number of factors. One was the Luftwaffe's dedication to providing ground support for the German army and thus its relative neglect of strategic targets, including the Soviet aviation industry. The Allies' Lend-Lease program was also helpful, and according to one recent Russian study, the United States and Great Britain provided the USSR with up to 18,303 aircraft during the war, but these planes began arriving in significant numbers only after the Soviet air services' recovery was well under way. Thus, that recovery was largely the result of Soviet measures; among these were the prewar policies outlined earlier and, once the war began, both the successful evacuation and reopening of aircraft and munitions plants and the creation of a talented command staff.

Fortunately for the Soviet Union, Stalin consistently supported these leaders' efforts to reorganize the air forces and their rear services. With his system of prewar command and control in ruins, he first followed the example of his tsarist predecessors by creating a powerful Headquarters of the Supreme Command on 23 June 1941 and a State Defense Committee on 30 June. Meanwhile, on 29 June his new *Stavka* appointed General P. V. Zhigarev to the new post of commander of the air forces of the Red Army. In order to rebuild and reorganize those services, Zhigarev immediately set out to correct the structural defects of frontal aviation, build up the reserves available to *Stavka,* and, in August 1941, set up a separate Rear Services Command under his own direction. In October he initiated operations against the Luftwaffe's airfields, and by early 1942, he felt strong enough to attempt an aerial blockade of the beleaguered German garrisons at Demiansk and Kholm. Even so, much remained to be done, as Zhigarev himself admitted in late January 1942. On 11 April he was replaced by his deputy, General A. A. Novikov.

Novikov retained his post throughout the war and is unquestionably one of the great commanders of aviation history. Under his aegis, the air force command carefully examined the lessons learned since June 1941. As a result, in May 1942 the service's capability for mass air operations began to improve. The structural overhaul was signaled by the transformation of the post–June 1941 frontal air forces into truly operational air armies. This involved the top-to-bottom restructuring of the air forces into air divisions, each comprising a single type of aircraft. Along with other reforms, this significantly raised the efficiency of the logistical support and command-and-control systems reinforcing the new and stronger frontline units. For example, in November 1942 the organizational table of fighter and ground attack regiments was set at three squadrons (twelve planes each) instead of two squadrons of three wings of three (nine planes in all).

In combat, direction of these stronger units was facilitated by the growing use of on-board radios, which allowed the aircrews to communicate with an expanded and denser network of ground-based early-warning and command-and-control stations. These last, along with improved navigational and weather services, were established for both the air and air defense forces from the summer of 1942 onward. Of equal importance were tactical changes. Apart from the creation of special groups of "aces," often designated as guards units, after September 1942, the "zonal" fighter patrols were supplemented by more autonomous "free hunters," or pairs of experienced pilots who roamed in search of enemy targets. And in December 1942, all these elements were integrated in the "aerial blockade" that Novikov (as a "representative" of

Stalin's *Stavka*) personally established to isolate the German Sixth Army within Fortress Stalingrad.

That the Red Air Force was able to rebound within eighteen months—going from a catastrophic defeat to besieging the Germans in Stalingrad—was due to the ability of the Soviet infrastructure to expand without Luftwaffe interference. There was a drop in production for 1941 while factories were moved east of the Urals, but then they turned out modern, heavily armed aircraft in such numbers that, by the end of the war, Soviet air armies deployed 2,500 to 3,000 machines each and were well supplied with munitions. At the same time, the requisite air and ground crews were provided in spite of losses and a vast expansion of new units. Numerical superiority was converted to effective command of the battlefields of *Stavka*'s choosing.

Conclusion

In both world wars, Russia's air services faced initial collapse but then recovered their capabilities to wage effective aerial combat. Needless to say, the Soviet effort of 1941–1945 was on a vastly grander scale than that mounted by the tsarist air services during 1914–1917. Although the recovery of the Red Air Force was more striking, its collapse in 1941 was more dramatic. But unlike the defeat of 1914–1915, 1941's failure cannot be even partially blamed on the fabled "backwardness" of Russian industry. Rather, it was Stalin's military purges and other immediate prewar policies that created the conditions that guaranteed the disaster that overwhelmed his air forces.

The same policies also did much to negate the overall progress resulting from the industrial modernization he had instituted after 1929. But in the long run, this process created both the industrial base and the pool of technically trained personnel that made the air forces' recovery possible. True, the Soviet dictator was exceptionally fortunate that both the aviation industry and the Red Air Force, like the other branches of the armed forces, contained reserves of talent that could replace that squandered by his purges. Despite the strategic errors of both Hitler and his overconfident and overextended Luftwaffe, the Soviet recovery would have been impossible without such leaders and designers, as well as the patriotic upsurge and courage of a multitude of average Russians.

This said, it must be admitted that this recovery would have been equally unlikely without Stalin's Five-Year Plans. In an effort to remedy the perceived weaknesses of the tsarist regime, these plans created the industrial plant and the correspondingly enlarged network of technical education and paramilitary training required for modern warfare. In the end, it was this base that provided Stalin with the ability to overcome the disastrous effects of his own domestic, military, and foreign politics and with the necessary preconditions for the resurrection of the air forces that he nearly lost in 1941.

Suggestions for Further Research

Although the wartime operations of the Red Air Force will continue to be reassessed (especially the role of Stalin) as new details emerge from the archives, a number of authorities have already discussed them in exhaustive detail. This is not the case, however, with regard to the period before 1917. Apart from that on the squadron of Sikorsky "Flying Ships," most serious research on the imperial air services is fairly recent, and much remains to be done. At present, as in the early days of flight, most of this work remains in the hands of gifted amateurs rather than those of academic scholars.

Two aspects of the Soviet air forces still await serious attention. One is the theme perused here: the training of technical and flying personnel through both the educational network and the "voluntary" societies supporting the air forces. Apart from William Odom's pioneering institutional study, the latter have yet to receive full-scale examination by a non-Soviet scholar, despite the obvious relevance to the more general concerns of underdevelopment and development in countries undergoing industrialization. The second neglected topic is the Soviet aircraft industry and, to an even greater extent, the aircraft armaments industry. Both await full-scale monographic treatments that incorporate post-Soviet and even Soviet assessments in a form accessible to non-Russian readers. Although histories of particular design bureaus have begun to emerge, we are still woefully ignorant of the early, crucial period, when the research and development system took root, and of the full extent of Stalin's impact on it.

Recommended Reading

Good introductions to the events discussed here can be found in Robert A. Kilmarx, *A History of Soviet Air Power* (London: Praeger, 1962); Von Hardesty, *Red Phoenix: The Rise of Soviet Air Power, 1941–1945* (Washington, D.C.: Smithsonian Institution, 1989); Robin Higham and Jacob Kipp, *Soviet Aviation and Air Power: A Historical View* (Boulder, Colo., and London: Westview and Brassey's, 1977, 1978), particularly the chapter "The Beginnings of Russian Air Power, 1907–1922" (pp. 15–34), which provides a background for World War I; and the more recent Robin Higham, John T. Greenwood, and Von Hardesty, eds., *Russian Aviation and Air Power in the Twentieth Century* (London: Frank Cass, 1998). As noted earlier, the only Western study devoted solely to the "voluntary" societies remains William E. Odom's *The Soviet Volunteers: Modernization and Bureaucracy in a Public Mass Organization* (Princeton, N.J.: Princeton University Press, 1973).

For other general histories and specialized studies on particular topics, readers are referred to the many entries with their extensive bibliographies in David R. Jones, ed., *The Military–Naval Encyclopedia of Russia and the Soviet Union/Military Encyclopedia of Russia and Eurasia* (Gulf Breeze, Fla.: Academic International Press, 1978–). In particular, consult the following: volume 1 (1978): "Ace, 1914–1917, Russian," pp. 138–40; "Aces, 1917–1945, Soviet," pp. 140–45; volume 5 (1994): "Aerial Armament," pp. 63–191; "Aerial Munitions," pp. 192–214; volume 6 (1996): "Aerial Blockade of Demiansk-Kholm, 1942–1943," pp. 115–200; volume 7 (1996): "Aerial Blockade of Stalingrad, 1942–1943," pp. 40–193; volume 8 (1998): "Aerial Bomb (Imperial Russia)," pp. 1–22; "Aerial Bomb (1918–1939)," pp. 23–51; and "Aerial

Bomb (1939–1945)," pp. 51–75. Finally, insight into German assessments of their Soviet opponents during 1941–1943 are provided by the study commissioned by the U.S. Air Force from former Luftwaffe general D. Walter Schwabedissen, *The Russian Air Force in the Eyes of German Commanders* (New York: Arno, 1968).

CHAPTER TEN

The United States in the Pacific

Mark Parillo

On 7 and 8 December 1941, aircraft of the Japanese navy and army conducted a series of devastating assaults on U.S. military targets in Hawaii and the Philippines as part of imperial Japan's first phase of operations against Allied forces in the Pacific. The Japanese onslaught dealt decisive defeats to American airpower in both locales, but there was a notable difference in the objectives of the two attacks.

The Pearl Harbor operation was a massive naval air raid designed to cripple the U.S. capacity to interfere with the Japanese seizure of economically and strategically vital territories all over East Asia and the western Pacific. The Japanese aerial attack on the Philippines was the prelude to an extended amphibious and ground campaign; hence the first day's strikes were only the opening moves of a sustained aerial offensive designed to gain control of the skies. The purpose and nature of the two operations were somewhat different, then, as were some of the reasons for the dual American defeats.

Pearl Harbor

The Pacific war began at 0755 on a sunny Sunday morning when the first Japanese bombs rained down on the U.S. Pacific Fleet anchored in Pearl Harbor, Oahu, Hawaii. Two principal waves of attacking aircraft, totaling 353 planes altogether, sank four battleships, seriously damaged four others, and destroyed ten other warships. It was a crippling blow to the U.S. Navy's surface strength in the Pacific, a force on whose deterrent value President Franklin Roosevelt had hung inflated hopes for preserving peace in the region.

The Japanese also singled out American airpower as a major target. The U.S. Army Air Forces at Wheeler, Hickam, and Bellows airfields; the U.S. Navy air bases at Kaneohe and Ford Island; and the U.S. Marine Corps air station at Ewa were all pounded, with most American aircraft caught on the ground. U.S. losses amounted to 188 aircraft destroyed and another 159 damaged of the 402 aircraft present when the raid began. Six B-17s and nine navy dive-bombers took off later in the day in a fruitless search for the attacking Japanese force—this meager retaliatory force was indicative of how enfeebled U.S. airpower in Hawaii had become after the day's stinging defeat. The Japanese had established air superiority over the islands for as long as their aircraft carriers cared to cruise the vicinity.

Raiding force commander Admiral Chuichi Nagumo, however, chose not to tempt fate by lingering in the area to launch additional strikes, so less than twelve hours after sending the first wave of attackers, the Japanese began their withdrawal. They left in their wake the smoking ruins of American warships, aircraft, and military pride (but not, as will be noted, fuel farms and workshops), and historians have devoted much time and energy in the decades since to examining the disaster. It is a complex story of miscommunication, lack of imagination, chance gone awry, incompetence, and, as some would have it, treachery.

The American defeat at Pearl Harbor was the result of several kinds of mistakes and inadequacies. Perhaps none of those failures has been more bemoaned and debated than the errors of American intelligence agencies, which might have provided warning of the raid, thus allowing better preparation of Hawaii's defenses and at least some mitigation of the massive losses suffered. Various conspiracy theories contend that intelligence methods, especially cryptanalysis, actually succeeded in ferreting out sufficient warning of the Japanese intentions but that American, British, or Dutch leaders, in the hope of uniting the United States and drawing it into the ongoing conflict, elected to withhold knowledge of the impending attack from those who needed it.

Those theories tend to founder on the logic of the situation. Had Roosevelt and other members of his administration known of the attack in advance, they would have been foolish to sacrifice one of the major instruments needed to win the war just to get the United States into it. Had another government received word that the United States was about to be attacked, American participation in the war would have been assured regardless of whether Washington knew of the attack in advance; withholding this vital intelligence only ran the risk of losing American trust along with the American military assets about to be destroyed.

It is unlikely, however, that the code breakers in the United States or any of its potential allies turned up definitive evidence of the Pearl Harbor raid before it happened. Documents declassified within the last ten years make it clear that more evidence of the attack was discovered via cryptanalysis than earlier historians had believed, including information that the Imperial Japanese Navy was practicing attacks on moored ships, outfitting special ordnance, and planning operations in the northern Pacific. Such intelligence fragments take on much more significance when viewed through the lens of historical hindsight, but at the time, they were tantalizing but ambiguous indications of what the Japanese might do—assuming that the tenuous diplomatic situation eventually resulted in war at all. Further, the most revealing fragments were scattered amid the dross of many thousands of other intelligence bits, some of which just as convincingly pointed to a Japanese attack on the Panama Canal and other operations that failed to materialize.

Although it is now possible to sift through the mounds of intercepted and decoded radio transmissions and find many arrows pointing to a surprise aerial assault on Pearl Harbor, contemporary intelligence operatives faced a far more daunting task. In addition to being confronted with the challenge of making sense out of a multitude of source materials of uneven quality, they were working in separate locations. There were U.S. Navy radio intercept and decoding stations in the Philippines and Hawaii as well as back in the continental United States. Theoretically "Cast," the station in the Philippines, and "Hypo," the code breakers at Pearl Harbor, sent all their materials to OP-20-GY in the U.S. Navy's offices at the Washington Navy Yard for comprehensive analysis, but in practice, the analysts relied as much on summary reports as on the actual intercepted messages. This substantially reduced the chances of a critical mass of the disparate intelligence clues about the Pearl Harbor raid ending up in the hands of an individual or small group who might have arrived at the right conclusions.

Still, in the hours before the outbreak of hostilities, cryptanalysis of Japanese diplomatic communications, transmitted in a code that was more readable than the principal imperial navy codes, alerted intelligence analysts back in Washington of the imminence of the breakdown of negotiations. The timing of the final Japanese diplomatic maneuvering suggested a possible attack on Hawaii, and Army Chief of Staff George C. Marshall dispatched a warning message to Major General Walter C. Short, the army commander in the islands, about an hour before the actual attack. The message was inexplicably sent through commercial channels, however, and arrived after the disaster.

There were other, more traditional means of detecting the attack beforehand that might have provided up to a couple of hours of precious time to ready the Pearl Harbor defenses. Radar was one possibility, and the Philippines and the Hawaiian department each had one set of radar in operation when they were attacked. But it was a relatively new technology, and radar was just being integrated into the aerial defense network at both places when the war began. The army's set in Hawaii was located on the heights at Opana Point to maximize reception, but there were as yet no communication lines, let alone standard procedures for reporting or evaluating the radar data, in place at the time of the attack. The soldiers manning the unit were still familiarizing themselves with the equipment when they detected a single aircraft approaching Oahu at 6:45 A.M. on the morning of 7 December, more than one hour before the first bombs fell. It was a scout plane from the attacking force, but the radar operators decided not to report it, assuming that it was an American aircraft. Fifteen minutes later the radar picked up a large body of aircraft bearing in from the northwest—Nagumo's first wave of aircraft. Their report to the air warning center at Fort Shafter, made from a public telephone some distance down the road from Opana, was dismissed as a flight of B-17s expected from the mainland.

Yet another tip-off of the looming attack was missed when the USS *Ward,* a destroyer patrolling the harbor entrance, sank one of the five Japanese midget submarines attempting to steal into the anchorage that morning, some hours after the little warship had first been spotted. The destroyer's notice of the incident reached Pacific Fleet headquarters more than an hour before the attack, but the message was marked low priority and was just being evaluated when the first bombers appeared overhead.

One might also legitimately wonder how the attacking forces—both the carrier task force on the day preceding the strike and the hundreds of aircraft that carried it out on Sunday morning—avoided detection by aerial reconnaissance of the surrounding waters, a fundamental defensive precaution for any major base. Both Admiral Husband E. Kimmel, commander in chief of the U.S. Pacific Fleet, and Rear Admiral Patrick N. L. Bellinger, commander of the naval air arm in Hawaii, recognized the need for aerial patrolling, especially of the islands' northern and western approaches. In March 1941 Bellinger had joined with his army counterpart, Brigadier General F. L. Martin, to pen a report to Washington detailing Pearl Harbor's vulnerability to air attack and noting the aircraft needed to patrol the area sufficiently: 180 B-17Ds. Since this was more than the entire U.S. Army Air Force inventory of B-17Ds at the time, there was scant chance of the desired aircraft being supplied anytime soon. Indeed, on the morning of the attack there were just six operational B-17s in Hawaii, and they were assigned to training missions, per instructions from Washington that emphasized the priority of training.

The onus of aerial patrol thus fell on the navy, which had three dozen PBY Catalina flying boats, well suited for such duties, available. Such numbers were inadequate for anything but partial coverage, but the problem was aggravated by shortages of personnel and maintenance spares, as well as the Navy Department's directive to prioritize training. In sum, Kimmel and Bellinger could provide thorough reconnaissance patrolling only for limited periods of time, and only by accepting that such a state would be followed by an extended period of downtime for servicing the hard-pressed aircraft. Given that the most recent war alert from Washington had been issued the weekend preceding the Japanese attack, Kimmel recognized the impossibility of maintaining full patrol schedules, lest he end up blinding his aerial eyes. Hence there were only three Catalinas in the air on the morning of 7 December; all the others were destroyed or heavily damaged in the attack on Kaneohe.

The failures of intelligence, radar, and aerial reconnaissance granted the Japanese the element of surprise, which reduced the effectiveness of U.S. defenses and consequently multiplied the destructiveness of the attack. Yet another factor exacerbating American unpreparedness on the day of the raid was the complicated command arrangements of the military forces in Hawaii and in general. By the terms of an old interservice agreement, the army and General Short were responsible for protecting the ships in Pearl Harbor. In addition, the naval command officially responsible for defense was not the commander in chief of the Pacific Fleet but the commander of the Fourteenth Naval District, a position filled in December 1941 by Admiral C. C. Bloch. Bloch was Kimmel's senior in rank, but Kimmel regularly micromanaged many of the details of Pearl Harbor's defense and otherwise intervened in Bloch's exercise of command. Some confusion and wasted effort inevitably resulted.

There was also a blurring of jurisdictional lines concerning intelligence within the War and Navy Departments. The service chiefs and secretaries withheld some potentially useful intelligence from Kimmel and Short, sometimes because they felt it was unnecessary to share it, and sometimes because they incorrectly assumed that the code breakers in Hawaii and in the Philippines already had access to it. For instance, the local commanders at Pearl Harbor never saw the "bomb plot" signals, intercepted Japanese consular messages asking for detailed grid maps of Pearl Harbor and the warship mooring locations.

It was not only intelligence matters that suffered because of the complex command and administrative setup. Direct Japanese–American negotiations had been going on since the spring, and Washington's instructions to the military commanders in Hawaii were not always clear and comprehensive. There had already been a couple of war scares before 7 December, and servicemen in the islands were growing complacent about being put on alert because nothing ever seemed to come of it. General Short in particular read the latest set of instructions, sent after the war scare of late November, to mean that the threat of war had abated to the point that the greater danger was now sabotage. Short informed Washington that he was taking precautions against sabotage and was never countermanded.

As a result, army aircraft were parked in tight rows out in the open to make them easier to guard from saboteurs, but this left them extremely vulnerable to air attack. Marine fighters and dive-bombers at Ewa were similarly arranged, ensuring heavy losses. The best defense of Pearl Harbor was probably the fifty P-40s and P-36s at Wheeler Field, but the meticulously parked planes—arranged with wingtips touching in rows twenty feet apart—were decimated by the first Japanese dive-bombing run.

The threat of sabotage had also led to reductions in the readiness of antiaircraft weapons. None of the navy's five-inch or larger guns and only 25 percent of the machine guns were manned, and the ready ammunition was secured in lockers to which only the duty officer had the key, although most guns were firing within fifteen minutes of the attack. The army's situation was worse, as only four of thirty-one antiaircraft batteries went into action at all. Also, given the reduced state of alert, one-third of the ships' captains were ashore, along with many of the other officers. It was, after all, Sunday.

On a more fundamental level, the American mind-set on the eve of the war was as much to blame as any specific problem of intelligence, equipment, or command. American racism, which would have infamous consequences for Japanese Americans in the years to come, led to an underestimation of the soon-to-be enemy. It was difficult for Americans to believe that the Japanese were capable of mounting such a complicated, long-distance operation. Reports from military and naval attachés, including those detailing the superiority of the nimble Zero naval fighter, were not read carefully. And, as happened elsewhere, information in the public domain, especially in newspapers and professional journals and commercial intelligence, was overlooked. The historical record, too, suggested the possibility of a Japanese surprise attack on the fleet: the Japanese had opened their victorious wars against the Chinese in 1894 and the Russians in 1904 with naval strikes before declaring war, the Royal Navy had changed the entire strategic situation in the Mediterranean in November 1940 with a carrier air strike against the Italian fleet at Taranto, and Western navies themselves had proved the feasibility of such attacks with an air strike on Singapore in 1938 during Royal Navy exercises and, on 7 February 1932, during U.S. Navy maneuvers, a Sunday-morning carrier raid on Pearl Harbor. Japanese–American relations had been worsening noticeably for years, and it was no secret that high-level negotiations were in trouble in the autumn of 1941. Any swabbie or dogface stationed in the Hawaiian islands, having recently undergone weeks of intermittent alerts and war scares, had to know that war was an imminent possibility.

It was not simply that the Japanese achieved tactical surprise, arriving in full force in the skies over Oahu virtually undetected; the initial stages of the attack were met as much by outright disbelief as by astonishment. Many of those present later recalled that their first reaction was the thought that the explosions were training exercises gone awry. Even after the feeling of unreality wore off over the next hours and days and weeks, many refused to believe that the Japanese had done it without German assistance, or even German participation, which was reported and sworn to by many witnesses, in and out of uniform. Others pointed an accusing finger at Japanese residents of the islands, believing that only widespread fifth-column activity could account for the stunning Japanese success. This sort of underestimation of Japanese capabilities bolstered the complacency and misjudgments that caused so many mistakes to be made and opportunities to be missed.

The Pearl Harbor raid was a stunning psychological blow to the American armed forces and people, and it was a tactical success of the first magnitude for the Japanese, who had lost less than thirty aircraft and a hundred men killed in the entire operation. Writers from Samuel Eliot Morison to today have decried the strategic error the Japanese committed by galvanizing the American people, but in truth, even the tactical military victory was a hollow one. The Japanese had raided Pearl Harbor to remove the threat to their offensive operations presented by the U.S. Pacific Fleet, and they achieved this to a greater degree and at a lower cost than they had anticipated. They had attacked other targets only to the extent necessary to strike a debilitating blow at the major naval units. The chance absence of the American carriers from the harbor on the morning of the attack preserved the U.S. Navy's principal offensive weapon in the Pacific. In addition, the Japanese had failed to recognize the ultimate value of some of the other available targets, such as the submarines, oil storage facilities, and major base installations. These proved to be as indispensable for the coming offensives as the fleet itself. The disabling of Pearl Harbor as a major fleet base would have forced the U.S. Pacific Fleet back to the American West Coast and delayed subsequent U.S. counterattacks in the Pacific by six months or longer. The defeat of American airpower in the space of a few short hours had rendered Pearl Harbor nearly, if only temporarily, defenseless, but Admiral Nagumo's decision to withdraw after only two strikes threw away some of the fruits of victory that courage and boldness had ripened on the vine.

For the Americans, the defeat was not as bad as it initially appeared. The sacrifice of more than 2,000 lives was certainly a tragedy, and the outlook surely would have been brighter without the heavy battleship and aircraft losses. But as the months ahead would continue to demonstrate, battleships could not operate under hostile skies, and aircraft were of little value without an adequate base and logistical infrastructure. Had the battle fleet somehow escaped the devastation of the Pearl

Harbor raid, it might have worried the Japanese as they advanced southward through the western Pacific, but it was unlikely to halt the waxing tide of the Rising Sun. In truth, the battleships might well have been irrevocably lost in waters other than the shallows of Pearl Harbor, from which six of the eight stricken capital ships were salvaged, repaired, and rearmed.

The American defeat on 7 December had another positive byproduct, which was the final resolution of the long-simmering debate over the value of naval aviation. The raid itself removed most doubts about the aircraft carrier's effectiveness, and the loss of so much of the navy's surface gunnery strength necessitated the rebuilding of the fleet's striking power around the surviving aircraft carriers. In this way, Pearl Harbor proved to be both a crushing American defeat and the first stone in the path to victory.

The Philippines

Just hours after the Pearl Harbor raid, as the morning sun began to climb the skies over the Philippines, the Japanese launched a second aerial onslaught, now aiming primarily at the archipelago's largest and northernmost island, Luzon. This resulted in a second defeat of American airpower, this time embodied in the U.S. Army's Far East Air Force (FEAF). More than one-third of the United States' most modern fighters and bombers—P-40Es, B-17Cs, and B-17Ds—were destroyed on the first day of the operation, a blow from which the FEAF never recovered. Available U.S. Navy air units consisted of little more than the thirty or so reconnaissance aircraft of Patrol Wing 10, mainly PBY Catalina flying boats, so the U.S. air defeat in the Philippines is largely the sad tale of the FEAF's demise.

The Japanese continued their offensive after the first day's successful strikes by launching a series of amphibious assaults beginning on 10 December, and eventually they overran the island. Despite the notable skill, perseverance, and courage displayed by American and Filipino troops in the prolonged resistance on Luzon's Bataan peninsula, within the first two weeks of operations, the ultimate outcome of the campaign was assured. The defeat of the FEAF in the war's opening round had doomed the Philippines and the Allied ground forces defending them to capture by the Japanese, a defeat more ignominious than Pearl Harbor because of its consequences on the ground.

Controversy has characterized the analysis of the American debacle almost from the moment it occurred. This is true of both the academic scholarship on the topic and the public dispute engaged in by some of the principal actors. Unquestionably, a fair portion of the controversy is bound up in the person of one of the chief protagonists, Douglas MacArthur, whose long and influential career has attracted unstinted praise and fierce excoriation in almost equal measure.

Outpost of empire and laboratory for American colonialism for four decades before World War II, the Philippines acquired a special strategic significance as tensions mounted in East Asia in the turbulent 1930s. Japanese expansion, brought on by the pressure for markets and materials generated by the nation's rapid industrialization in the late nineteenth century, had led inevitably to clashes with the Western powers, which dominated much of the region's trade and consequently had the most to lose. Utilizing the mutual distrust of the Great Powers to fullest advantage up to and including World War I, Japan had boosted its influence on the Asian mainland and in the rich territories of the western Pacific. This had been accomplished through both force of arms and diplomacy. Wars with China and Russia resulted in the acquisition of Korea as a colonial possession and growing influence in Manchuria, while Japanese intervention in World War I on the Allied side provided the basis for seizure of German territories in China and the Pacific, which Japan retained directly or administered on behalf of the newly formed League of Nations. The 1920s witnessed the trial of the "new diplomacy" of multilateralism in reaction to the perceived dangers of covert bilateralism, and Japan's foreign policy seemed to conform to the new order by its signing of international agreements, such as the 1922 Washington treaties and Kellogg-Briand Pact of 1928, designed to stabilize the world order by limiting armaments, promoting commercial opportunities, and erecting nonviolent alternatives to war for the settlement of international disputes. Specifically, the Washington treaties sought to make Japan feel secure by not allowing a major Royal Navy base closer than Singapore or a U.S. Navy base nearer than Pearl Harbor.

The onset of a severe economic crisis in the industrialized world in the early 1930s spelled the doom of multilateralism, however, and when the Japanese military eliminated or intimidated their rivals for control of national policy in the same period, the stage was set for a renewed employment of force. In 1931 the Imperial Japanese Army began an outright conquest of Manchuria, followed by the political domination of additional provinces in north China and then, in 1937, the outbreak of open warfare against the government of Nationalist China. Continued Japanese aggression in China led to mounting criticism, followed by economic pressure, from the Western powers, particularly Great Britain and the United States, and Japanese policy makers began to seriously consider military action to break the nation's economic dependence on imports from the West. The Pearl Harbor raid had been designed to remove the U.S. Pacific Fleet from the campaigns of Japanese conquest, but the base itself held far less significance in Japanese strategic thinking.

The Philippines, however, assumed a greater strategic role in the minds of all concerned. The islands did not contain vital resources themselves, but their location was directly astride the only shipping lanes between the locus of Japan's industrial center in the Home Islands and the richly endowed lands to the south. In particular, the Japanese economy and military forces required petroleum, of which there was precious little within the Japanese empire. But the Netherlands East Indies, some 2,000 miles due south of the Japanese Home Islands, produced 10 million tons of crude oil annually, which amounted to 50 percent more than Japan needed. Seizure of the Dutch colony would end the empire's single most pronounced vulnerability to economic pressure, and war planners could not contemplate waging a campaign for its capture, not to mention securing the routes for the vital commodity's shipment to Japan, without gaining dominion over the Philippines.

U.S. planners, pondering since 1898 the challenges of defending an island mass 7,000 miles distant from the North American continent, had often disagreed on whether the Philippines could be held at all in the event of hostilities with Japan. The role of the Philippines in strategic planning, from War Plan Orange to the later Rainbow Plans, had varied over the years as weapons and military doctrine changed, and in the final months before the commencement of hostilities in the Pacific, those plans, particularly the role envisioned for the aerial component of the defense forces, changed again.

The first aircraft for defense of the islands appeared in 1920, but until the eve of World War II, only obsolescent U.S. Army designs could be found there. In 1940, while Spitfires and Me 109Es were tangling over the British Isles at speeds in excess of 350 miles per hour, the frontline fighter plane in the Philippines was the P-26A, a fixed-undercarriage monoplane with a top speed of less than 240 miles per hour. The only bombers available were seventeen twin-engine Martin B-10Bs, originally designed in 1932 and no longer produced after 1936. Observation and reconnaissance aircraft were older, even antiquated. Compounding the problem of inferior aircraft were other shortcomings, such as inadequate numbers of pilots and aircrews, insufficient training of air personnel assigned to the islands, airfields of uneven quality and insufficient quantity, lack of any modern detection and warning system, and obsolete antiaircraft guns.

The reasons for the neglect of the Philippine air defense contingent were twofold: strategic mission and budget constraints. An Air Board report from September 1939 described the defense forces in the Philippines as expendable because the islands could not be held. Given the chronically limited appropriations for defense during the Great Depression, the Philippines, which was positioned well down the feeding trough from the congressional slop chute, received slim sustenance. By 1941, the penury of the Far East Air Force was a well-established tradition.

The tale of the FEAF's defeat in the early days of the Pacific war is much more than the demise of a neglected stepchild, however, because in the months before the Japanese attack, the mission and nature of the

FEAF underwent a transformation. The reasons for this were both military and political.

In the autumn of 1940, as the FEAF faced the daunting prospect of withstanding a potential invasion of the Philippines with its paltry allotment of aging equipment and undertrained personnel, the Royal Air Force's gallant Fighter Command was blunting the first sustained aerial offensive of World War II in the Battle of Britain. Air Chief Marshal Sir Hugh Dowding, commander in chief of Fighter Command during the crucial phase of the battle, arrived in the United States in January 1941 on behalf of the British Purchasing Commission. War Department officials consulted Dowding about the feasibility of mounting an aerial defense against an invasion of the Philippines—something War Department planner Captain Hoyt S. Vandenberg had been advocating since 1939—and Dowding expressed confidence that with modern fighter aircraft and some bombers to disable the invaders' transports, the Philippines could be held. Although some U.S. Army officers agreed, Chief of Staff George C. Marshall did not, and plans for upgrading and expanding the fighter defenses of the Philippines met with his disapproval. Marshall did, however, authorize the dispatch of limited numbers of the army's most modern fighter, the P-40B (eventually followed by the improved P-40E), and some B-18 medium bombers to replace the archaic B-10s.

It was not the presence of a modern fighter but the operational readiness of the B-17 Flying Fortress, however, that triggered a volte-face in U.S. policy concerning the Philippines. An early version of the B-17 had been shipped to England and reached operational status in the summer of 1941, and the performance of the first few dozen of these big four-engine bombers was greatly overestimated by some in the War Department, including Secretary of War Henry L. Stimson. This, combined with Marshall's reprioritizing of the Japanese threat after their acquisition of additional bases in French Indochina, was enough to convince President Franklin D. Roosevelt to base a force of B-17s in the Philippines as a deterrent to further Japanese aggression in the western Pacific. In August 1941 the War Department initiated plans for a B-17 ferry route to the Philippines, hoping to have more than 150 there by the spring of 1942, with more of the Fortresses plus some of the new B-24s (with capabilities generally similar to the B-17's) to follow. New pilots and aircrews also arrived, and engineers began to upgrade facilities, most notably by lengthening Del Monte airfield on the southern Philippine island of Mindanao to handle the big bombers.

By the time of the Japanese attack in December 1941, the FEAF was much improved from just twelve months earlier. It could call on about ninety fighters (P-40Bs, P-40Es, and P-35As), thirty-five B-17Cs and B-17Ds, and fifteen B-18s, plus assorted scout planes. New construction was under way, albeit slowly. Although there were still staff, training, and equipment deficiencies, as Lewis Brereton, the newly appointed FEAF commander, recalled after his initial inspection tour in early November 1941, "While everyone was suffering from lack of equipment, encouraging steps were being made to prepare the units for active service." By April 1942, the islands would have nearly 200 four-engine bombers, assorted light bombers (including some dive-bombers, which would be invaluable for attacking enemy troop convoys), and 250 modern fighters available for their defense, with sufficient base infrastructure, adequately trained personnel, and a radar-based warning system.

The Japanese offensive in December caught the FEAF in the midst of the buildup process, but at least the Americans were considerably better prepared than if the attack had come just a few months earlier. The Japanese had more than 400 bombers and fighters engaged in the campaign, including 90 of the excellent Zero fighters. Operating from bases in Formosa 300 miles to the north, Imperial Japanese Navy bombers and Zeroes savaged Clark and Iba Fields on Luzon, catching many American aircraft on the ground and destroying more than 100 planes. Twelve of the burning hulks were B-17s, and several more of the heavy bombers were damaged. The attacks also knocked out the only operating radar set in the islands. The remaining B-17s mounted several strikes against attacking forces within the next couple of days, scoring very poorly against Japanese ships in attacks carried out at altitudes varying from 7,000 to 25,000 feet. After two days, the big bombers withdrew to Del Monte airfield 600 miles to the south, and by 21 December, the fourteen that were still operational had all flown back to Australia. Although some of them returned a few days later for a final round of attacks against the invaders, all surviving American aircraft, including the dwindling number of fighters, received evacuation orders on 24 December and departed for Australia. The Japanese, who had anticipated the loss of up to 40 percent of their aircraft during the campaign, had suffered far lower losses and now roamed the skies over the Philippines at will. The FEAF had suffered a stinging defeat.

Despite its recent reinforcement and upgrading before the war, the FEAF faced steep odds when hostilities commenced. The Japanese had superior numbers, and their aircrews had much more extensive training and combat experience from the ongoing war in China. The Japanese also had sound aircraft designs, numerous and adequately equipped bases, and, given the Imperial Japanese Navy's command of the waters surrounding the Philippines, a far better strategic position. A successful defense of the islands was probably impossible under those conditions.

Nevertheless, the FEAF had the resources and the opportunities to make the Japanese conquest more costly. The B-17s provided some genuine offensive punch, and amphibious operations (risky under the best of circumstances) often exposed the attackers—particularly the slow and vulnerable transports and supporting vessels needed to storm a hostile shore—to counterstrokes from a determined defender. A bold and active defense repulsed the first Japanese expedition sent to take Wake Island, for instance, and a small U.S. destroyer force caused a panic at Balikpapan, Borneo, when it torpedoed several Japanese transports in a bold raid on 24 January. Given the forces at hand, the FEAF might have accomplished more than it did in America's first sustained aerial operations of the Pacific war.

There were several reasons for the FEAF's performance failures. Especially damaging to the FEAF's long-range prospects were the heavy losses on the first day of the war, consisting largely of aircraft destroyed on the ground some nine hours after the attack on Pearl Harbor. Although informed of the Pearl Harbor raid hours before daylight and the first Japanese strikes in the Philippines, the FEAF was caught with most of its aircraft still on the ground, which accounts for the heavy losses. Variable weather and the long distances involved—Formosa is hundreds of miles to the north and in a different climatic zone—caused delays and uncertainties about launching air operations, and the failure to pass on visual sightings and radar warnings of approaching Japanese aircraft resulted in a debacle.

In postwar interviews and writings, Douglas MacArthur, commander of the U.S. Army Forces in the Far East, and his principal air commander, Lewis Brereton, exchanged accusations of responsibility for the disaster. To this day, it is not clear why MacArthur was not more energetic about getting the B-17s off the ground on the morning of 8 December, and the most recent appraisals hold him primarily responsible, though Brereton shares some of the blame. Historian William H. Bartsch has also noted the role of lesser officers in not relaying warning messages about the impending Japanese attack to Clark Field.

The tragic loss of so much fighting strength on the ground in the first day's action, followed by years of wrangling over the personal responsibility of the various individuals in the chain of command, has focused too much attention on the short-term reasons for the FEAF's defeat. Brereton himself later ascribed the critical loss of airpower to a violation of the fundamental principles of security and mobility—that is, the inadequacies of the FEAF infrastructure at the time of the attack: a rudimentary warning and communications system (including only one operating radar set, which was lost on 8 December), shortages of modern antiaircraft weaponry, and the paucity of airfields, especially those with runways long enough for the B-17s.

There is much truth to this assertion, for despite the recent change in its strategic role, the FEAF was in the midst of a buildup and refurbishment when the Japanese attack materialized. Modern aircraft and trained personnel were only beginning to arrive in significant numbers. For example, the pilots for four squadrons of dive-bombers debarked in Manila on 20 November, but their aircraft had not arrived by the time of the attack and were eventually diverted to Australia. Ammunition for the .50-caliber machine guns of the new P-40Es was in such short supply that many of the guns had not been test-fired when the actual shooting war started; adequate quantities of ammunition would not even be leaving the United States until March 1942. Personnel and equipment of all sorts were still arriving and undergoing training and acclimatization. Facilities, from runways to communications equipment, were undergoing expansion and thus labored at only partial capacity. Brereton's staff recommended that aircraft engines be shipped back to the continental United States for overhauling until the planned fiftyfold increase in engine maintenance capacity could be completed. The FEAF was very much a force in transition when the war broke out, months away from any reasonable semblance of battle readiness.

There is also reason to believe that the loss of one-third of the main aerial offensive force, the B-17s, was not as calamitous as it appeared at the time and has frequently been depicted since. American military and government leaders had great faith in the B-17's destructive potential, and much of the acrimony in the postwar recriminations about the FEAF's unpreparedness rested on the conviction of Brereton, MacArthur, and others that the heavy bombers, if wisely applied, could have effectively disrupted the enemy assault. The Japanese themselves were concerned that the news of the Pearl Harbor raid would trigger damaging U.S. air attacks on their bases in Formosa as they completed their final preparations for offensive operations.

These expectations about the B-17's performance were, however, based on insufficient practical experience. The B-17's debut in Europe had been disappointing at first, and modifications in design, tactics, and procedures were required before the aircraft came into its own as a strategic bomber later in the war. More pertinent to the Philippines situation was the ineffectiveness of high-level bombing against ships at sea, as seen in the few antishipping strikes the surviving B-17s launched over the next two weeks and reconfirmed six months later during the Battle of Midway. The bombers might have been used for tactical support or interdiction missions against Japanese forces on Luzon once the land campaign was under way, but the poor bombing accuracy at that stage of the war, combined with the heavy ground cover, the enemy's talent for camouflage, and the four-engine aircraft's performance characteristics, rendered these possible uses of the B-17 less than promising. Indeed, the later Pacific war saw little use of heavy bombers for ground support roles, except for their attacks on bridges or other transportation targets. The B-17s' most worthwhile employment in the Philippines probably would have been attacking enemy air and naval bases. Assuming that the Japanese air defenses could have been penetrated, a sustained aerial offensive by the American five-squadron heavy bomber contingent might have inflicted enough damage to dislocate Japanese plans to some degree.

The B-17s of the FEAF, however, lacked the infrastructure necessary to undertake continuous operations without incurring rapid attrition of forces. Aside from a meager bomb supply and the scarcity of suitable bases—only Clark and Del Monte airfields could handle the big bombers—there was a severe shortage of tools, spare parts, and engines (not "so much as an extra washer or nut for the Flying Fortresses"). The B-17 in the Philippines in 1941 was only a shadow of the weapon it would become in later years; in fact, it was not even the weapon it could have been in 1941. Even if the unfortunate losses of the first day had been avoided, the FEAF's prospects for withstanding the Japanese offensive against the islands were relatively slim.

Force structure inadequacies, command failures, overstretched resources, and unfounded optimism in U.S. aircraft designs were the most significant immediate causes of the FEAF's defeat in December 1941. Yet these causes were themselves the products of deeper and more fundamental weaknesses of the U.S. air forces at the outset of the Pacific war, even aside from the already discussed underestimation of the Japanese opponent. One of these weaknesses was the American tradition of neglecting defense in peacetime, only to be caught unprepared when international problems erupted into war. This trend had long debilitated America's peacetime establishment, but it was especially problematic in 1941 because of the particular conditions of the preceding decade, and it affected the defense of Hawaii as well as the Philippines.

The United States, like the rest of the industrialized world, suffered through the most serious and prolonged economic crisis of modern times, and defense appropriations took a backseat to the momentous task of regenerating the economy. The Roosevelt administration had little choice but to funnel energy and resources toward combating the Great Depression, even as the greatest conflict in human history began to unfold. Despite some diversion of funds to naval construction—some cruisers were built with money from the New Deal's National Recovery Administration, for instance—the armed forces were confronted with bare-bones budgets for years, forcing them to make impossible choices about prioritizing missions and programs. The FEAF's story of neglect may have been more pathetic than that of many other commands or agencies, but it was a difference of degree only. Neither the Philippines nor Hawaii was going to receive the forces the military deemed necessary for their defense while so many millions of Americans lived in poverty or its shadow.

Adding to the problem was the growing complexity of warfare and military forces in the industrial age. Improvisation and breakneck mobilization might suffice to stave off catastrophe in the nineteenth century, and it might even work in the early twentieth century, when the United States could rely on its oceanic insulation to buy time, but the rapid development of airpower had greatly reduced the margin of safety by the time of World War II. Weapons and other equipment were more intricate and required more resources to produce, and they were needed in massive quantities. People, too, were required in greater numbers and with more specialized training than ever before. Airpower was beginning to bridge the oceans, so the time available for mobilization was shrinking even as the task of defense grew larger and more complex. Japan, already fighting for years before the Pacific war began, did not face the same challenge of emergency mobilization.

The defense of the Philippines in particular suffered from another handicap. Widespread overconfidence in the efficacy of strategic bombing, not confined to American policy makers, led to the FEAF being placed in a position of considerable diplomatic importance as the force charged with intimidating the Japanese into changing their national policy. The problem was that such a position also made the FEAF a prime target for Japanese military action, just as the presence of the U.S. Pacific Fleet had prompted a Japanese attack on Pearl Harbor. And until the FEAF was actually built up into the force that the Roosevelt administration hoped to wield on the stage of Pacific diplomacy, it was not only vulnerable to attack but also all the more likely to be assailed. Thus, one can argue that the FEAF was as much a victim of American diplomacy as it was of the feeble economy.

Conclusion

American policies on the eve of the Pacific war may seem foolish in hindsight, but in fact, the defeat of the U.S. air forces in the Pacific was also emblematic of some of the strengths that would lead to eventual U.S. victory. Even if there had been room in the budget for rearmament in the 1930s, such a program would have been politically suicidal for any administration during the Great Depression. The diplomacy that failed to stop war from coming to the Pacific also led to the war's eruption through a Japanese act of unequivocal aggression, which helped unite Americans and gird their will for the long process of winning a mass-production war. In the end, the U.S. air forces in the Philippines and Hawaii were defeated because they were asked to do the impossible, the result of subordinating military needs and plans to national policy. This was a far healthier and, ultimately, more militarily effective course than the reverse—as the Japanese would learn in dramatic and unforgettable fashion.

Suggestions for Further Research

Perhaps no American military experience has received more popular and scholarly attention than the opening battles of World War II in the Pacific, yet there are important dimensions of those events that remain unexplored. Many of the reasons behind Japan's victories lay in logistics, both the Japanese proficiency at it and the U.S. shortcomings. It is striking that Japanese forces projected greater aerial and naval power in Hawaii and the Philippines yet did not deploy greater numbers of aircraft than the defenders, who had occupied those areas for two generations, yet there is no systematic or comprehensive study of how and why this state of affairs was achieved.

Another promising avenue of inquiry involves the transition in military technology that was occurring at the time of the attacks: radar surveillance was being implemented but was not yet operational; code breaking was entering the age of growing reliance on mathematical and mechanical solutions, as well as confronting the problems of analyzing mushrooming collections of data; and aircraft design and support systems were at a revolutionary stage.

One might also explore the U.S. Pacific disasters in terms of the intellectual transitions in military thought and planning that were coincident with the technological transition, including the growing belief in the efficacy of aerial defense, the political and psychological value of airpower and strategic bombing capabilities in particular, and the challenges to the long-esteemed Mahanian strategic tradition posed by naval aviation and submarines.

A balanced reassessment might also answer the question: Should the United States have been surprised? Would a study of naval building policies and practices, as portrayed in the works of Norman Friedman, be revealing?

Scholars might also investigate the forces—theory, education, military experience, cultural values—that shaped the Japanese and American strategic Weltanschauung that produced the performances of influential individuals, from Nagumo's reluctance to launch additional strikes on Pearl Harbor to MacArthur's judgment concerning the timing and employment of his available airpower.

Recommended Reading

Given the shock to the national psyche caused by the Japanese attack on Pearl Harbor, one should not be surprised at the vastness of the literature devoted to the opening battles of the Pacific war. One would do well to begin with official histories, particularly Wesley Frank Caven and James Lea Cate, *The Army Air Forces in World War II,* vol. 1, *Plans and Early Operations, January 1939 to August 1942* (Chicago: University of Chicago Press, 1948), and Samuel Eliot Morison, *History of United States Naval Operations in World War II,* vol. 3, *The Rising Sun in the Pacific, 1931–April 1942* (Boston: Little, Brown, 1948). More comprehensive accounts of various aspects of the operations have been published in the decades since these volumes appeared, but they remain valuable sources from which the student can gain a solid overall grounding in the events before moving on to more specialized studies.

There are also numerous general histories of World War II and of the Pacific theater. Williamson Murray and Allan R. Millett, *A War to Be Won: Fighting the Second World War* (Cambridge: Harvard University Press, 2000); Ronald H. Spector, *Eagle against the Sun: The American War with Japan* (New York: Free Press, 1985); and John Costello, *The Pacific War* (New York: Rawson, Wade, 1981), are among the most comprehensive and balanced accounts. H. P. Willmott provides great insight into both Japanese and American strategic thinking in *The Great Crusade: A New Complete History of the Second World War* (New York: Free Press, 1989) and the more topically focused *Empires in the Balance: Japanese and Allied Pacific Strategies to April 1942* (Annapolis, Md.: Naval Institute Press, 1982). To these one might add some overall histories of the Japanese navy and Japanese naval aviation: David C. Evans and Mark R. Peattie, *Kaigun: Strategy, Tactics, and Technology in the Imperial Japanese Navy, 1887–1941* (Annapolis, Md.: Naval Institute Press, 1997); Paul S. Dull, *A Battle History of the Imperial Japanese Navy, 1941–1945* (Annapolis, Md.: Naval Institute Press, 1978); and Masatake Okumiya and Jiro Horikoshi, with Martin Caidin, *Zero!* (New York: E. P. Dutton, 1956).

The historiography of Pearl Harbor is inextricably linked with the traditional–revisionist debate over conspiracy theories. The traditional view, that the Pacific war was the result of conflicting policies and tragic misjudgments, was first fully laid out in Herbert Feis, *The Road to Pearl Harbor: The Coming of the War between the United States and Japan* (Princeton, N.J.: Princeton University Press, 1950), but since American code-breaking successes have been made public, an entire literature has been devoted to the questions of who knew what and when they knew it. Roberta Wohlstetter, in *Pearl Harbor: Warning and Decision* (Stanford, Calif.: Stanford University Press, 1962), was the first to conclude that the intelligence gleaned from cryptanalysis was too scattered and indistinct among the background "noise" to provide sufficient warning under normal circumstances, and that theme has been reaffirmed and developed in several later works: David Kahn, *The Codebreakers: The Story of Secret Writing* (New York: Scribner, 1967, 1996); Ladislas Farago, *Broken Seal: The Story of Operation Magic and the Pearl Harbor Disaster* (New York: Random House, 1967); Gordon Prange, with Donald M. Goldstein and Katherine V. Dillon, *At Dawn We Slept: The Untold Story of Pearl Harbor* (New York: McGraw-Hill, 1981); Edwin T. Layton, with Roger Pineau and John Costello, *And I Was There: Pearl Harbor and Midway—Breaking the Secrets* (New York: William Morrow, 1985); Frederick D. Parker, *A Priceless Advantage: U.S. Navy Communications Intelligence and the Battles of the Coral Sea, Midway, and the Aleutians* (Washington, D.C.: National Security Agency, 1994); and John Prados, *Combined Fleet Decoded: The Secret History of American Intelligence and the Japanese Navy in World War II* (New York: Random House, 1995). Presenting variations on the revisionist theories, such as that radio intelligence provided forewarning of the attack but was purposely suppressed for political reasons, are John Toland, *Infamy: Pearl Harbor and Its Aftermath* (Garden City, N.Y.: Doubleday, 1982); James Rusbridger and Eric Nave, *Betrayal at Pearl Harbor: How Churchill Lured Roosevelt into World War II* (Old Tappan, N.J.: Simon and Schuster, 1992); John Costello, *Days of Infamy: MacArthur, Roosevelt, Churchill, the Shocking Truth Revealed: How Their Secret Deals and Strategic Blunders Caused Disasters at Pearl Harbor and the Philippines* (New York: Pocket Books, 1994); Robert B. Stinnett, *Day of Deceit: The Truth about FDR and Pearl Harbor* (New York: Free Press, 1999); and Timothy

Wilford, *Pearl Harbor Redefined: USN Radio Intelligence in 1941* (Lanham, Md.: University Press of America, 2001).

There are a few general histories of the Pearl Harbor attack that provide reliable overall narratives of the event, including Gordon W. Prange, with Donald M. Goldstein and Katherine V. Dillon, *December 7, 1941: The Day the Japanese Attacked Pearl Harbor* (New York: McGraw-Hill, 1987); Archie Satterfield, *The Day the War Began* (Westport, Conn.: Praeger, 1992); Michael Slackman, *Target: Pearl Harbor* (Honolulu: University of Hawaii, 1990); and Stanley Weintraub, *Long Day's Journey into War: December 7, 1941* (New York: Penguin, 1991). Also of interest is Clark G. Reynolds, *The Fast Carriers: The Forging of an Air Navy* (New York: McGraw-Hill, 1968), which puts the raid into the larger context of World War II naval aviation.

For a more comprehensive picture of the Pearl Harbor tragedy and the reasons for it, one would do well to look at the memoirs, biographies, and autobiographies of the event's principal actors, such as Hiroyuki Agawa, *The Reluctant Admiral: Yamamoto and the Imperial Japanese Navy* (Tokyo: Kodansha, 1965; New York: Hill and Wang, 1994); Henry C. Clausen and Bruce Lee, *Pearl Harbor: Final Judgment* (New York: Crown 1992); Donald M. Goldstein and Katherine V. Dillon, eds., *Fading Victory: The Diary of Admiral Matome Ugaki, 1941–1945* (Pittsburgh: Pittsburgh University Press, 1991); and Husband E. Kimmel, *Admiral Kimmel's Story* (Chicago: Regnery, 1955). A spirited defense of two of the principals in the attack can be found in Edward L. Beach, *Scapegoats: A Defense of Kimmel and Short at Pearl Harbor* (Annapolis, Md.: Naval Institute Press, 1995). The stories of the lesser-known participants can be found in Paul Stillwell, ed., *Air Raid, Pearl Harbor! Recollections of a Day of Infamy* (Annapolis, Md.: Naval Institute Press, 1981), and Robert S. LaForte and Ronald E. Marcello, eds., *Remembering Pearl Harbor: Eyewitness Accounts by U.S. Military Men and Women* (New York: Scholarly Resources, 1991).

For more comprehensive coverage of the Pearl Harbor literature, see Myron J. Smith Jr., *Pearl Harbor, 1941: A Bibliography* (New York: Greenwood, 1991), and Eugene L. Rasor, "The Japanese Attack on Pearl

Harbor," in *World War II in Asia and the Pacific and the War's Aftermath, with General Themes: A Handbook of Literature and Research,* ed. Loyd E. Lee (Westport, Conn.: Greenwood, 1998).

The history of the American aerial defeat in the Philippines is in some ways a more complex story than the one-day raid on Pearl Harbor. A firm basis for further study is contained in the U.S. Army official history: Louis Morton, *The Fall of the Philippines* (Washington, D.C.: Office of the Chief of Military History, U.S. Army, 1953). The Japanese side can be found in three of the volumes of the official history series from the Boeicho Boekenshusho Senshishitsu [Japanese Defense Agency, Military History Section], *Senshi Sosho* (Tokyo: Asagumo Shimbunsha): vol. 2, *Hito Koryaku Sakusen* [Philippine Invasion Operations] (1966); vol. 24, *Hito Malay Homen Kaigun Shinko Sakusen* [Malayan Offensive Army Air Operations] (1969); and vol. 34, *Nampo Shinko Rikugun Koku Sakusen* [Southern Offensive Army Air Operations] (1970).

Some of the earliest campaign histories are Robert L. Scott, *Damned to Glory* (New York: Charles Scribner' Sons, 1944); Robert Reynolds, *Of Mice and Men* (Philadelphia: Dorrance, 1947); and Walter D. Edmonds, *They Fought with What They Had* (Boston: Little, Brown, 1951). Later, more comprehensive accounts include John Toland, *But Not in Shame* (New York: Random House, 1961); Ted Williams, *Rogues from Bataan* (New York: Carlton Press, 1970); James Leutze, *A Different Kind of Victory* (Annapolis, Md.: Naval Institute Press, 1981); and Dorothy Cave, *Beyond Courage* (Las Cruces, N.M.: Yucca Tree Press, 1992). Many of these campaign histories focus much more on the heroic defense of Bataan than on the opening operations or the air war. Some excellent recent accounts, however, have rectified this gap in the literature: Richard Connaughton, *MacArthur and Defeat in the Philippines* (New York: Overlook, 2001); William Bartsch, *Doomed at the Start* (College Station: Texas A&M University Press, 1992); and Bartsch's definitive *December 8, 1941: MacArthur's Pearl Harbor* (College Station: Texas A&M University Press, 2003).

For a sense of the Philippines' place in prewar American planning, see Edward S. Miller, *War Plan Orange: The U.S. Strategy to Defeat Japan, 1897–1945* (Annapolis, Md.: Naval Institute Press, 1991); Brian McAllister Linn, *Guardians of Empire: The U.S. Army and the Pacific, 1902–1940* (Chapel Hill: University of North Carolina Press, 1997); and James C. Gaston, *Planning the American Air War: Four Men and Nine Days in 1941* (Washington, D.C.: National Defense University Press, 1982). Numerous biographies and recollections are pertinent to the topic, too, including Lewis H. Brereton, *The Brereton Diaries* (New York: William Morrow, 1946); Henry H. Arnold, *Global Mission* (New York: Harper, 1949); John J. Beck, *MacArthur and Wainwright: Sacrifice of the Philippines* (Albuquerque: University of New Mexico Press, 1974); D. Clayton James, *The Years of MacArthur,* vol. 2, *1941–1945* (Boston: Houghton Mifflin, 1975); William Manchester, *American Caesar* (New York: Little, Brown, 1978); Paul P. Rogers, *The Good Years: MacArthur and Sutherland* (New York: Praeger, 1990); Forrest C. Pogue, *George C. Marshall: Ordeal and Hope* (New York: Viking, 1966); and Philip S. Meilinger, *Hoyt S. Vandenberg: The Life of a General* (Bloomington: Indiana University Press, 1989).

CHAPTER ELEVEN

Defeats of the Royal Air Force

Norway, France, Greece, and Malaya, 1940–1942

Robin Higham and Stephen J. Harris

During the Second World War, the British Royal Air Force (RAF) suffered four resounding defeats—the air campaigns in Norway, France, Greece, and Malaya—and achieved one decisive victory—the Battle of Britain—before the U.S. Army Air Forces began to make a significant contribution. One other campaign, the fighter offensive over France in 1941–1942, was neither a victory nor a defeat, although it cost Fighter Command dearly.

The roots of these defeats, as well as the one victory, lay in the interwar years and reflected a combination of political, economic, military, technical, and psychological factors. The political and economic are easiest to document. Although the RAF had emerged from the Great War as the second largest (but probably most powerful) air force in the world, its strength was rapidly dissipated in a postwar world where France was the only visible (though unlikely) threat to the United Kingdom and Britain itself was the only world power. That fact meant that the defense of the empire could not be altogether ignored, but at home, the logic of the Ten-Year Rule introduced in 1919 was unimpeachable: because there would be no major war for a decade, the services need not plan or equip themselves for such a conflict. Treasury control was paramount, but it was reinforced by a foreign policy (and diplomats) that preferred to see a pacific future and ignored, for example, Japan's attitude after the 1921–1922 rupture of the Anglo-Japanese alliance or the darker side of German demands that the harsh terms of the Treaty of Versailles be softened.

For their part, the service chiefs never failed to warn their government that disarmament (or lack of rearmament) was being carried too far, but it was only in the early 1930s—indeed, after the disappointing failure of the 1934 Geneva Conference—that their arguments began to resonate along Downing Street. British rearmament began, fitfully, the next year, but almost two decades of cuts had had their effect. There was much to do in what turned out to be only four short years, and priorities had to be established.

So far as the RAF was concerned, having successfully fought the navy and army to maintain its independence, Chief of the Air Staff Sir Hugh (later Lord) Trenchard was able to produce a cadre service established on a sound technical footing; however, his interest in grand strategy, beyond mere assertions that aerial bombardment could destroy an enemy's will (or means) to fight, was minimal. Without much critical analysis, then, the corollary to this assertion—that aerial bombing was likely to be decisive—shaped the organization of the RAF, so that the ratio of bombers to fighters in Britain was two to one. Whether the air force could do something useful to support the army—something with which the RAF (and Trenchard) had considerable experience during the First World War—was not ignored entirely, but the organization responsible, Army Co-operation Command, was a poor relation to Fighter and Bomber Commands when the RAF embraced its functional structure in 1936.

With its emphasis on offensive bombing, it followed that when the existence of the new German Luftwaffe was revealed in 1935, the RAF and the politicians it advised mistook the former's raison d'etre completely. The German air force had been planned as a tactical force to assist a continental army to win land battles, but in Britain, it was seen as a mirror image of the RAF—a "strategic" air force whose primary objective was wiping out London. Consequently, following the personal intervention of the minister for coordination of defense—who, among other things, had to end the squabble between Hugh Dowding's Fighter Command and the bomber barons who dominated the RAF—the air defense role and fighters got priority. Army Co-operation Command, already on the fringes, was isolated even further, as was the need to think about providing air support to an army involved in mobile warfare.

The decision to give priority to air defense happened to coincide with the first fruits of the technical-technological revolution in aircraft design, production, and operations that, along with breakthroughs in radar development, actually made effective air defense possible. The year 1936 saw the first flights of the new monoplane fighters, the Hurricane and the Spitfire, with their Rolls-Royce Merlin engines and eight machine guns in the wings. At the same time, real heavy bombers were ordered off the drawing boards to be in service by 1942. The exponential jump in the money and man-hours required to design and build these planes, and the requirement for new airfields, caused the Air Ministry's budget to pass £450 million in 1939 (up from £16 million in 1932), but the simple expenditure of money could not hasten the appearance of a modern and robust RAF. Technological developments took time, as did the shift of public opinion from peace at any price to the inevitability of war with Germany. At the time of the Munich crisis (September 1938)—and when war began a year later—the RAF had medium and heavy bombers in service that could strike the Ruhr but not Berlin, although the capital was regarded as the decisive target, where cataclysmic effects were possible. Within Fighter Command, meanwhile, the closing out of biplane production and the rush to bring Hurricanes and Spitfires into squadron service led to a situation in which there was no inventory of spare parts to effect wartime repairs, and there were too few trained mechanics. Available means, in short, did not support desired ends.

A potential "force multiplier" existed for the RAF in the form of the French Armée de l'Air, and indeed, there had been desultory talks between London and Paris about air matters from 1936 to 1938, but these bore only middling fruit. For one thing, the French air force was about two years behind the RAF technically and financially, and it was probably no less confused when it came to understanding what the Luftwaffe was meant to be. Moreover, London and Paris were only loosely "allied" and at times had very different perceptions of what the coming war would be like. One critical difference was in the area of air defense. While Britain was making progress, and Dowding could conceive of a unified and coordinated system using observers, radar, and sector control, the French—so much nearer to Germany—had little prospect of being able to defend their cities and were therefore predisposed against the bombing of German cities, lest that provoke a German response.

By September 1939, it could be said that the Allies had agreed on a broad general strategy. A British Expeditionary Force would go to France with integral air support, but its ability to move forward to meet the anticipated German thrust through the Low Countries could not be definitively arranged until Belgium, ardently neutral until May 1940, agreed. Where British aircraft would be based and how they would be used to support the fighting on the Western Front were even more nebulous questions with equally nebulous answers, and in this regard, there was certainly no shared doctrine between the British and French air forces. While acknowledging the usefulness of bombers in *support* of an advancing army, for example, the British air staff was not convinced that using a bomber force *against* an advancing army, well supported by antiaircraft defenses and fighter aircraft, would be economically effective. But that is what the French proposed.

In the Far East, meanwhile, there was even less of a British grand strategy, despite rising Japanese ambitions. This was not for want of trying on London's part, at least in the years immediately following the First World War, but the 1919 Jellicoe plan calling for Canadian, Anzac, and British collaboration in the event of a Pacific war against Japan had grounded on the shoals of Dominion reluctance and could not be refloated. British policy consequently became dogmatically defensive—and dilatory—concentrating on fortifying the naval base at Singapore from which the Main Fleet would operate.

The common flaw in the European and Far Eastern strategies was the failure of intelligence—in part due to parsimony and, to a greater degree, to lack of languages—along with the lack of knowledge of and inability to think like the potential opponent. Indeed, until the March 1941 Singleton review, the Air Ministry consistently credited the Luftwaffe with thousands more aircraft than its actual strength. Such exaggeration led British planners astray in terms of both German capabilities and intentions.

So, neither in Europe in 1939–1940 nor in the Far East in 1941–1942 were the British prepared mentally or physically for the war that engulfed them.

Norway

Norway came first, and from the Allied perspective, it was meant to come first in a preemptive attempt to block German access to Narvik iron ore. But the Allied effort was preempted by the Germans, who seized Oslo, Stavanger, Trondheim, and Narvik first. Intervening in both the center and the north, the Anglo-French response was wholly inadequate. "No air forces need to accompany the . . . army in the first instance," the British chiefs of staff ruled when it came to central Norway, although they allowed that the air contingent planned for their original Narvik operation might reinforce success. But a decision on that could safely be "deferred." In the event, one squadron of Gladiators, the last of the biplanes, was dispatched from HMS *Glorious.* They had to use a frozen lake as a makeshift base, causing their engines to freeze, and ten of eighteen were destroyed on the ground when the Germans attacked. Four more were shot down. Subsequently, all air support in the central region was provided by U.K.-based machines, with twin-engine Blenheims having to take on the fighter role.

Prospects were better around Narvik, which was farther from the Luftwaffe's recently captured airfields, and two squadrons, one each of Gladiators and Hurricanes, were sent. Both were able to intercept German air attacks and to take on German ground forces, to good effect. Indeed, there was success here worthy of reinforcement. But once the enemy had launched attacks on Holland, Belgium, and France on 10

May, there was no prospect of sending more aircraft: Bomber Command's twenty-three squadrons were earmarked for attacks on Germany proper, while Fighter Command and the RAF in France would stay where they were. Consequently, no matter how gallant, the British effort in Norway was doomed, and after successfully covering the army's withdrawal, the surviving fighters were flown off to the *Glorious.* They sank with the ship.

German audacity, luck, and competence turned the Anglo-French Norwegian campaign into a lost cause that depleted stores that were urgently required elsewhere. The special British handicap was the need to improvise airfields (and the delivery of supplies), whereas the Germans could exploit essentially interior lines and use the existing Norwegian infrastructure. That, and the Allied inferiority in the air, was the main reason why the RAF and other services felt the full effects of the Luftwaffe's power. That the Germans lost more aircraft (242) than the RAF (29) may have been testimony to the leavening effect of far-flung expeditionary operations on the Luftwaffe and the vulnerability of transport aircraft (accounting for about eighty of the German planes lost). Nevertheless, for the RAF, Norway was a classic example of a three-dimensional campaign fought with too little, too late; it was too far-ranging and based on inadequate local intelligence. There was too little hard, thoughtful consideration in London; the optimistic British planners overestimated the efficacy of both antiaircraft artillery and air raids on enemy airfields. They also behaved as if political or operational objectives were administratively possible, leading to impulsive, ad hoc decisions at the highest levels that paid no attention to practicalities at lower levels—or to the realities of logistics. Such practices may have been inevitable, especially given the dispatch of most experienced staff officers to the buildup in France, but they were reinforced by the absence of a supreme commander or combined headquarters to plan and direct the affair. Norway, in short, was a sideshow that, in retrospect, should have been a no-show. The conclusion of the RAF's initial historian, that the importance of air superiority, "if not new as a theory . . . was new as a fact" after Norway, beggars belief and must be dismissed as bunk. Yet, something of that mentality was evident in the campaign that was under way to the south, in the heart of western Europe.

France

That Germany was a potential common enemy was clear to at least some in London and Paris as early as March 1936, when Adolf Hitler ordered his army into the Rhineland. Yet serious talks between the British and French military staffs did not begin until three years later, after Germany had swallowed up what remained of Czechoslovakia. These discussions resulted in a plan based on the static defenses of the Maginot Line; an advance on the left into Holland, to allow a defensive battle to be fought far forward of France's industrial north; and light forces covering the Ardennes, where the French high command dismissed the possibility of a strong German thrust.

France expected to mobilize 86 divisions (including 12 on the Italian frontier), while the "Phoney War" following the German victory in Poland allowed the British Expeditionary Force (BEF) to grow to about 10 divisions, giving the two allies a fighting chance against the 116 divisions the Germans had in the west—at least in terms of numbers. The Dutch and Belgian armies would help even the balance, but execution of the plan depended on Belgium recanting its traditional neutrality, which did not happen until May 1940. That forced the BEF into an unplanned battle near the Dyle River, but the Germans' Ardennes offensive was the main focus—and it was there that the Allied defenses unraveled irrevocably.

From the RAF's perspective, the air battle in France was shaped by the overall collapse on the ground, but the air campaign had weaknesses of its own. Forces dedicated to a direct attack on the German army were limited. Most of Bomber Command remained in England, where doctrine called for it to attack German industry, not the German advance (although the French, for various reasons, preferred the latter); army cooperation was not in Bomber Command's playbook. Most of Fighter Command would also be withheld from France—a hedge against failure there. The dedicated BEF Air Component, then, was a mixed bag of fighter, reconnaissance, and light bomber squadrons based on French airfields but lacking transport, radar, and any form of organized fighter sector control. The Advanced Air Striking Force (AASF), essentially Bomber Command's No. 2 Group with fighter squadrons attached, amounted to just over ten squadrons. Stationed in the French sector of the front, behind the Maginot Line, it was theoretically capable of taking on "strategic" targets in western Germany or choke points on the enemy's lines of communication—enemy concentrations west of the Rhine and rail yards east of the river.

The British contribution gave the Allies about 2,600 aircraft against Germany's 3,700 and Italy's 1,000. Numbers counted, but so did qualitative differences. The single-engine Fairey Battles of the Air Component and AASF could not survive daylight operations; the twin-engine Blenheims were usable but vulnerable; the Lysanders were good enough for artillery spotting but not for intelligence at the operational level. The Hurricanes were a match for any German fighter then in service, but there were too few of them to do much good. (The French air force, for its part, simply could not man and maintain all the modern aircraft in its inventory.) Many of the British airfields were within striking range of the Luftwaffe, and because the French antiaircraft arm was not strong, they were poorly served by ground defenses. Some, seeded with New Zealand grass, would not be fully serviceable until midsummer. Moreover, owing to the lack of trucks and other transport, units at threatened or inadequate fields could not easily transfer operations elsewhere; flying off aircraft was one thing, but relocating ground crews, fuel, armaments, spares, and repair machinery was another.

Air Marshal R. S. Barratt was in overall command of the British air forces; Air Vice-Marshal C. H. R. Blount led the BEF Air Component, and Air Vice-Marshal P. H. L. Playfair the AASF. Overall command in the northeast was vested in General Georges of France, and Barratt's headquarters was wisely colocated with Georges's and that of General d'Astier de la Vigerie, commander of French air forces in that theater. Together they could issue orders to the AASF but could only make requests for Bomber Command support from England. Their command-and-control task was aided by the Allied Central Air Bureau and the high-powered wireless telegraph listening posts of the special Hopkinson mission.

The Germans struck nine of the AASF's ten airfields on 10 May, as well as railways and supply points, in their effort to cripple the Allied air forces, but the only real damage done was to Barratt's headquarters, where the communications facilities were destroyed. These attacks cost the Luftwaffe about half its planes—daylight sorties were risky for everyone. For their part, RAF offensive sorties were able to cover Lord Gort's advance to the Dyle—although that advance played into German hands—but by day's end, the Fairey Battle force sent to Luxembourg had been decimated. Indeed, by 12 May, the AASF had lost 63 of its 135 bombers, and the chief of the air staff informed Barratt that he must husband his resources. This was easier decreed than done. The German breakthrough on the Meuse the next day meant that the AASF could not be held back, and on the fourteenth, trying desperately to help the French near Sedan, Playfair lost 40 of the 71 aircraft he dispatched. Two days later, rolling the survivors of ten squadrons into six, the AASF withdrew to the south—first to Troyes, then to Le Mans—in an effort to stay out of reach of the advancing Germans while still being able to intervene and not outrun its own supplies. Playfair continued to dispatch attack sorties, but they had marginal effect, and on 15 June the remainder of his bomber aircraft were sent home. The fighters returned to Britain a few days later after covering the withdrawal of the three British divisions left in France. Although the AASF flew bravely, under great stress and with little rest, a succession of uncoordinated and serendipitous local successes failed to influence the course of the land battle as a whole in any appreciable way, and 229 aircraft were lost in the process. Lacking any early-warning system, the fighter squadrons meant to protect AASF airfields were essentially impotent, and if they tried to escort the bombers to their targets, they were often overwhelmed by superior German numbers. Those Blenheim crews who managed to make it to their objectives were then endangered by their own faulty tactics, flying at 12,000 feet (rather than on the deck) and making multiple passes over targets that were well guarded by quad 20 mm flak.

In the north, the RAF Air Component suffered a similar fate, but sooner, and it probably did even less than the AASF to affect land operations. Its Fairey Battles were shot out of the sky during the day, and the quick German advance soon threatened its bases at Amiens and Abbeville. Indeed by 19 May, just nine days after the campaign began, the Air Component was withdrawn to Britain; the Germans were too close, and their forces could now be hit from British bases. Of the 261 Hurricanes that had been dispatched, 66 returned to Britain, 75 were written off, and 120 damaged machines were left in France.

More British aircraft in France would have made some difference—but how significant, and at what cost? Although aware of the risks—the airfields in France were not secure—Prime Minister (and Minister of Defense) Winston Churchill approved the release of thirty-two additional Hurricanes on 13 May—Spitfires were never going to be sent—and the next day he urged that ten additional squadrons be dispatched. Persuaded that there were too few servicing units available—and that Fighter Command's strength must not be diluted too far—he initially agreed that none should go, but full of emotion, he raised the issue again. Four more squadrons were sent to the Continent, but when France asked for six more on the sixteenth, the air staff insisted that their transfer could not be permanent; instead, three at a time would be flown to French airfields for half a day only. Churchill agreed, but the German advance was so quick that this procedure lasted only three days. By then, the gap between the BEF in the north and the main French armies to the south was deemed unclosable, and the Air Component began to return to the United Kingdom.

Of course, the intention had never been to move the main strength of Bomber Command to France. What was at issue was whether it should strike targets associated with the German advance or try to carry the war to the Ruhr no matter how pressing the former became. Adhering to prewar thinking and plans, and following the Luftwaffe's bombing of Rotterdam, which removed any moral constraints, the choice was to attack Germany. Thus began what would become an almost five-year offensive, the effects of which are still argued. But it is certain that these operations had little if any impact on the course of the war in France and Belgium.

Once it was obvious that the BEF would have to be extracted from France, the focus of RAF operations became clearer. From 26 May until the last troops embarked, Bomber Command mounted 651 sorties and Fighter Command 2,739 in support of the Dunkirk evacuation, and although the troops on the ground were rarely satisfied with what they saw, the effect of the air support there was much greater than in the fighting withdrawal from the Dyle. Still, the Battle of France was a decisive defeat for the RAF because it failed to come anywhere near achieving its admittedly vague objectives. Intended to support a more or less static front, it was totally out of its depth, given the mobile nature of the campaign that developed, for which it was not designed, equipped, or organized. When communications broke down, forward airfields were overrun, and contact with the front was lost, the British air forces in France were left to flail ineffectually with aircraft unsuitable for their tasks (the exception being the Hurricane). There was nothing in the British inventory—or British doctrine—to replicate the dive-bombing of the Ju 87 or the close coordination that existed between the army and the *Geschwadern* of Heinkels, Dorniers, and Messerschmitts. Still, the German operations did not come cheaply, and the Luftwaffe lost 1,284 aircraft of all types from all causes in the west. That the RAF lost only 959 was due to one factor: it had committed so few aircraft. Yet when its air forces were employed as prewar thought and policy had foreseen and prepared for, with all the associated radar, ground observers, and sector control and from bases safe from German armored forces, Fighter Command was able to win the Battle of Britain.

Greece

Defending the United Kingdom remained the obvious strategic imperative after the fall of France, but Britain also had to secure Suez Canal and the lifeline to India and the Far East. Italy was the initial opponent here, moving north and east from Abyssinia in June 1940, and then east from Libya into Egypt in September—the front and side doors to the canal area. Both offensives were blunted, not crushed, by the time Italy invaded Greece through Albania in October, a potential back door to the Middle East. Secretary of War Anthony Eden argued that British air forces in the Mediterranean were too weak (just twenty-nine squadrons, with 350 Gladiators, Blenheims, Lysanders, and Sunderlands and 400 or so truly ancient machines) to fight on three fronts and help the Greeks—the last people still fighting the Axis on the Continent. Air Chief Marshal Sir Arthur Longmore decided that he must do something, however, and dispatched a mixed squadron of Blenheim fighters and bombers to strengthen the air defenses of Athens. The government agreed, approving three more squadrons—two of Blenheims, one of Gladiators—as well as a squadron's worth of the latter for the Greek air force.

Conditions in Greece may have been less trying than those encountered in Norway a few months earlier, but even so, the RAF had once again sent an expeditionary force abroad with little understanding of what it was getting into. Disease was rampant, sanitation and mess facilities were primitive, and the Greeks themselves were denying access to the best airfield sites near Salonika for fear of upsetting the Germans, who now occupied Bulgaria. It was another case of make-do, especially for the bomber squadrons, which had to operate from grass fields with poor drainage 300 miles from the front.

There is no doubt that the Blenheim and Gladiator fighters were valuable additions to the Greek air defenses, although it certainly helped that they were fighting the Italians, not the Germans. Whether British offensive operations contributed significantly is debatable, however. True to his training and to RAF doctrine, the local commander, Air Vice-Marshal J. H. D'Albiac, resisted Greek efforts to persuade him to provide close air support to their army until February 1941, when a thrown-together wing of fighters and bombers based on dry patches of ground near Paramythia did good work in support of a Greek offensive into Albania. But once the Italians counterattacked, D'Albiac withdrew these aircraft, contending that their role there was not the "proper employment of an air force."

By then, RAF strength had grown to seven squadrons, but it was anticipated that the Germans might enter the fray soon. Having decided to stand with the Greeks—four divisions were to be sent from the Middle East—London provided two additional squadrons, bringing the total to nine, and undertook to replace more Gladiators with Hurricanes. The Germans struck on 6 April 1941 from Bulgaria and Yugoslavia with twenty-seven divisions (seven of them armored) supported by more than a thousand aircraft. Facing them were six Greek, twenty-four Yugoslav, and three (soon to be four) British and Commonwealth (Anzac) divisions; a hundred or so Yugoslav and Greek aircraft; and the nine RAF squadrons. Additional Greek forces and two more RAF squadrons continued to battle the Italians on the Albanian frontier. With these odds and the Germans' qualitative superiority over the Greek and Yugoslav armies, the outcome was never in doubt. Salonika fell on 8 April, and within a week, the Luftwaffe (and the lack of repair facilities) had reduced the effective strength of D'Albiac's northern force to 46 aircraft. When weather permitted, they had achieved some good results against the Germans' communication lines and marching columns, but as in France, once the armies they were supporting began to collapse, they shifted roles to cover the evacuation of British and Imperial forces from the mainland. By then, however, there were simply too few aircraft left to protect themselves; on 23 April, for example, 13 Hurricanes were destroyed on the ground. In the end, staging through Crete, only 57 of the 255 planes dispatched returned to Egypt.

The RAF commitment to Greece faced considerable odds prior to Germany's involvement, and impossible ones thereafter. The Gladiators were a match for the Italian CR.42 but less so for the Macchi G.50, and they were completely outclassed by the Luftwaffe's Messerschmitt 109. The Blenheims used in the fighter role were adequate, as was the bomber when the weather allowed. (The Wellingtons sent from the

Middle East on temporary deployment were the true top of the line). But too often, RAF crews found themselves having to operate from inconvenient (and not very robust) grass fields that were susceptible to flooding, had a weak maintenance and repair organization, and were not covered by any modern air defense system. Moreover, there was little opportunity for reinforcements once the Germans were involved, because the Mediterranean theater itself was short of aircraft. As always, the soldiers complained about the RAF's reluctance (and inability) to support them, and there was some truth to the charge: it was not part of doctrine, and D'Albiac was only too glad to return to more conventional operations when the Italians counterattacked out of Albania in February. Still, the blame attached to General Archibald Wavell, the army's Middle East commander in chief, and to Air Chief Marshal Longmore, his RAF counterpart, seems harsh. Their forces were defeated in a campaign imposed on them by London in a "noble" effort to assist a doomed ally; in fact, the Greek leader Metaxas had always said that his country would collapse if the Germans invaded. Although the army and air force did well enough against the Italians, they became forlorn hopes against the Germans. The RAF could have survived only if London had ordered its withdrawal on 6 April.

The Far East: Malaya and Burma

Malaya is a long peninsula, not as mountainous as Greece, but covered with tropical jungle and many streams that hinder north–south travel. Rainfall is heavy throughout the year, and there are frequent violent thunderstorms, with heavy, low clouds that were impenetrable to the aircraft of the day. Kuantan is 300 miles south of Singora, which is roughly 500 miles north of Singapore. There were 18,000 Europeans living in Malaya in 1939, of whom 42 percent were women and children. This population was reduced in 1941 to 9,000, nearly all of whom were in government service. This was a cumbersome structure unsuited to war and certainly did not have the makings of a pool of reliable, local labor.

Having failed for almost two decades to appreciate Japanese ambitions, neither London nor the local authorities had done much to improve or modernize the region's defenses against that empire, apart from those guarding the seaward approaches to Singapore. In a way, this was understandable: none of Britain's traditional imperial rivals would have been expected to attack overland through the Malay states, and Japanese capabilities to do so were seriously underestimated. Moreover, the difficult terrain, lack of roads, and long coastline suggested that the defense of the mainland was a task primarily for the air force rather than the army. The general policy of the Far East Command was therefore to construct as many airfields as possible, grouped to allow the concentration of the aircraft expected to be available. Twenty-six were completed by December 1941 on both coasts, but of these, fifteen were grass fields that became treacherously muddy in the wet climate. Two, at Kota Baharu and Kuantan, were located in close proximity to good invasion beaches, where they would immediately fall into enemy hands given a successful assault. Improvements to the grass fields could not be made because of the small labor pool; most "native" workers were employed in the tin mines and rubber plantations. In addition, overcoming the drainage problem would have required heavy equipment that was simply not available. Furthermore, although everyone knew that the airfields needed their own air defense artillery, by December 1941, only 17 percent of that required had reached Malaya, and most of the forward, vulnerable airfields had no antiaircraft defenses at all.

Air defense was also critical to the security of Singapore. Although four radar stations had been installed there and an observer corps created, there were gaps in coverage, and the lack of telephone lines limited their effectiveness. Efforts to promote civil defenses on the island were hampered by the water table; blackout was considered impossible due to ventilation needs; and although enough food supplies for 5 million people to survive a six-month-long siege had been imported, they were stored around Alor Setar, in the northwest, and would likely fall into enemy hands almost immediately. In short, from the standpoint of managing and caring for the civilian population, the island was not a firm base.

At the outbreak of war on 8 December 1941, there were four daytime fighter squadrons and one night squadron, limited to the defense of Singapore. In addition, there were two light bomber, two general reconnaissance, two torpedo bomber, and one flying boat squadrons—a combination of RAF, Royal Australian Air Force, and Royal New Zealand Air Force units, with Dutch air force reinforcements (twenty-two Glenn Curtiss bombers and nine Brewster Buffaloes) promised. There were only eighty-eight aircraft in reserve. Of the newly arrived Buffaloes (each of which needed twenty-seven modifications), only three squadrons were ready operationally. Not all the pilots were well trained when hostilities began, nor, due to unprocessed intelligence, did they know that the enemy's Zeros were faster. The other types of aircraft available were old and, in the case of the Vildebeeste torpedo bombers (two squadrons), long overdue for replacement. The light bombers (the four Blenheim I and IV squadrons) and the GR Hudsons (two Australian squadrons) were trained for overseas work but lacked reserves and had too many possible roles. Most important, the number of aircraft available was far below the 330 the British chiefs of staff considered minimal and the 566 requested.

Given this shortfall, primary responsibility for defense was suddenly switched to the army, which meant that the airfields became a defense liability and a danger if taken by the enemy. They had been prepared for demolition by sinking concrete cylinders into the runways and then adding metal canisters of explosives, but these would prove to be less than effective measures when the airfields were overrun. Meanwhile, repairing bomb craters on the runways before the Japanese arrived was an endless, nearly impossible task, given the water table and the continuing lack of native labor. Conscription of the civilian population was never implemented.

The army was not prepared for this unexpected role. Trained for neither jungle nor mobile warfare, many of its 87,000 personnel were afraid of the ground they would have to fight over. Yet, at the same time, they were sublimely overconfident and led largely by officers who did not appreciate the lessons of France and the Middle East and underrated the Japanese. (Overall command was vested in Air Chief Marshal Sir Robert Brooke-Popham, who handed control over to Lieutenant General Sir Henry Pownall on 23 December 1941. In an attempt to produce a coordinated theater effort, General Sir Archibald Wavell became the supreme commander of the American, British, Dutch, and Australian area on 29 December. The air officer, commanding [AOC] was Air Vice-Marshal C. W. H. Pulford; he became ill and was eventually replaced by Air Vice-Marshal P. C. Maltby. The army commander in Malaya was Lieutenant General A. E. Percival.) The high command was faced with a Hobson's choice: it could meet the Japanese landings well forward and defend the airfields, or, in anticipation of landings at Mersing (close to Singapore) or even on the island itself, it could hold forces farther to the south. The former option was chosen, but the ground commander in the north had too few forces in an area unsuited to a strategic and tactical defense. It went without saying that the two sets of airfields had to be defended, but his lines of communication were limited to a single-line railway. Thinking broadly and imaginatively, the high command conceived of Operation Matador, a preemptive action into Siam; planning for this operation so absorbed the staff that little attention could be paid to the defense of the airfields. In theory, at least, preemption was one way around the army's weakness—it might catch the Japanese off balance—but since it involved a violation of Thai neutrality, Matador required advance authorization, and on 5 December London told the governor and the commander in chief that Matador could be launched only after the Japanese had landed in Siam or in the Dutch East Indies. By then, it would be too late, as the initiative would be in enemy hands. And that is what happened. Despite the warnings of impending Japanese operations, political concerns about Thai neutrality led to the cancellation of Matador on 7 December.

During the night of 7–8 December, three experienced divisions of the Japanese army landed both north and south of the Thai–Malay border, with one regiment intending to push on into Burma. Their objective from the beginning was to seize the British airfields in northern Malaya as well as that at Victoria Point, Burma—the first step in their attempt to gain overall air superiority in the theater, as well as to cut off the reinforcement and resupply chain from India. Singapore was also attacked in the early morning. Radar gave adequate warning, and the antiaircraft defenses went to immediate readiness (the searchlight units were slower to react). Although there was no blackout and the moon was full, damage at the main targets, the Tengh and Selatar airfields, was slight.

Night fighters were not scrambled against the Singapore raid because they had not practiced with the air defense organization, but there was an active response against the invasion force reported at Kota Baharu. Indeed, an initial attack by the Australian Hudsons was successful, sinking one transport, damaging two others, and killing as many as 3,000 Japanese; a second attack, this time by Vildebeestes, arrived after most of the Japanese transports were gone. The pilots therefore chose to land at Kedah and Kelantan airfields to refuel, but they were caught on the ground by the Japanese, and most of their aircraft were destroyed.

As in France, the nature of the British air campaign was now profoundly shaped by events on the ground. Having failed to stop the amphibious landing, the army could not keep the Japanese from the fringes of Kota Baharu airfield, and the five Hudsons and seven Vildebeestes there withdrew to Kuantan. Other airfields in the north were hit before the Allies could bomb the landings at Singora and Patani, and all but two Blenheims were destroyed on the ground at Alor Setar. With only 50 of 110 aircraft still serviceable in the north, it was time to withdraw and implement the airfield denial scheme, but the demolitions were incomplete, leaving the runways intact. At little loss to themselves, the Japanese were well on their way to establishing air superiority.

The AOC now decided to use his depleted bomber forces against the newly established Japanese base at Singora—the most dangerous of all the Japanese landings. Six Blenheims, lacking fighter support, attacked and lost three of their number; the Buffaloes at Butterworth on the west coast were fully engaged in standing patrols over that airfield. Another attempt, this time with fighter escorts, was clobbered by high-level bombers and low-level fighters before it got off the ground. Two days later, on 10 December, Admiral Tom Phillips's attempt to intervene against the Japanese landings without air cover met with disaster, as the capital ships *Prince of Wales* and *Repulse* were sunk, lending additional credibility to the enemy's aura of invincibility. The Butterworth base was evacuated, with only eight aircraft having survived. The AOC tried to reestablish a network of fields in central and southern Malaya using improvised administrative and maintenance arrangements and whatever stores had been successfully transported from the north by rail. Even so, priority had to be given to the defense of incoming convoys.

The battle continued to go poorly on all fronts. The Indian army's Eleventh Division, for example, was driven out of its position by just two Japanese battalions and a company of tanks. Although air reinforcements were arriving—six Hudsons and five Blenheims by Christmas Day, and fifty-one crated Hurricanes early in the new year—the lack of air transport into Singapore meant that there was a shortage of servicing personnel. And when they did arrive, superior Japanese forces took their toll. Although some of the Hurricane reinforcements had success on 20 January against an unescorted Japanese raid on Singapore, the next day, five fell victim to the escorting Zeros, shot down at low level, where the Hurricane's performance was decidedly inferior.

The Allied air forces' inevitable defeat was hastened as the Japanese army continued to overrun abandoned airfields that had not been fully demolished—and the radar sites. The loss of early warning made life difficult for the Hurricane pilots and practically impossible for the Buffalo squadrons, which needed more than thirty minutes to reach incoming bombers at their normal 25,000 feet. At the same time, offensive operations against the Japanese landings continued to be costly: thirteen Vildebeestes were lost in two attacks on the Japanese landings at Endan on 26 January. Their efforts were praised by the army commander, who noted that they had proceeded "unflinchingly to almost certain death in obsolete aircraft which should have been replaced many years before."

The Japanese success at Endan rendered all further defense of the Malay Peninsula impossible, and the army began to withdraw into the island fortress. With the airfields there under attack, his bomber force practically written off, and only twenty-seven fighters left—twenty-one Hurricanes and six Buffaloes—Pulford knew that the battle was lost. On 27 January he ordered his remaining bombers to Sumatra, and Wavell concurred that only eight fighters should remain to defend Singapore itself. The last aircraft flew off on 10 February, and the island was surrendered five days later. The air force continued the fight on Sumatra until it too became untenable; most survivors withdrew to Java, and a smaller number escaped to Australia.

Ever since 1918, but particularly after 1922, Singapore and Malaya had been far distant places less known in Britain than even Czechoslovakia had been in 1938. Authorities in the prosperous and vital region were commerce-minded and did not wish to be disturbed by such practicalities as preparing for defense—an issue on which, in the event, the three services could not agree. Even as Japanese power grew, the area seemed secure until the fall of France, when French Indochina ceased to be a barrier and British forces had to be concentrated nearer to home. That fact that no powerful (and charismatic) generalissimo was appointed, along with the shortage of trained staff officers, reduced the potential to do more with less.

The Japanese occupation of southern Indochina in July 1941, which gave them access to airfields within striking distance of Singapore, might have been a final indicator that more had to be done to boost the region's ability to defend itself, but the misreading of Japanese intentions (it seems that despite U.S.-led provocations, everyone everywhere believed that the enemy would strike somewhere else) meant that wishful thinking (and poor staff work) prevailed. The likely course of Japanese action was missed, and no intelligence network to transmit enemy progress in the north to Singapore, 500 miles away, was established. After the two capital ships were lost and the air forces were withdrawn south, the Japanese army had a much freer hand. Allied troops were aware that everything was crumbling and began disintegrating themselves, both physically and psychologically, too often abandoning their positions to inferior attacking forces. Suffering the fate of a colonial outpost, receiving the dregs of equipment and the least-experienced officers, staff, and pilots, the Allied air force in Southeast Asia became essentially irrelevant, its operations no more than a pinprick. Malaya fell in seventy days because its only real chance for survival had been an absence of war in the Far East.

As many as 400 aircraft supported Japan's 1941 thrust into Burma near Mandalay, where they would initially be opposed by 16 RAF Buffaloes, 41 Curtiss P-40s of Claire Chennault's American Volunteer Group, and some horribly outdated de Havilland Moths of the Burmese air force. Infrastructure in Burma was actually much better than in Malaya, and reinforcements did arrive: a Blenheim squadron and 30 Hurricanes just after Christmas. Furthermore, the AOC there demonstrated a strong offensive spirit, ordering his crews to lean forward against the enemy. In the beginning, the Allies won temporary air superiority over Rangoon, but in the end, the Japanese army prevailed, taking over their airfields. The Allied squadrons engaged in a "fighting withdrawal" as they made their way back to India, but on 27 March a Japanese attack essentially wiped out all that was left of their aircraft. Personnel retired by rail and road into India and China. The Japanese, meanwhile, were now within range of Calcutta. It was never bombed, but the Indian east coast and the huge British base at Trincomalee, Ceylon, were attacked in April.

Conclusion

The doctrinal underpinning of British air operations in Norway and France was a triumph of hope over the capabilities of current technology, augmented by a complete failure to understand the basic infrastructure requirements of deployed squadrons. Thereafter, the air staff assumed that the RAF's victory in the Battle of Britain could be repeated overseas against first-class air forces without the necessary radar control and communications, top-of-the-line fighters, and logistics. All this was inadequate to counter the enemy's operational plans. RAF planning was almost useless because London approached the task with the assumption that it knew how the enemy would act. There was no sense of the enemy's *Schwerpunk,* or vital aim. The result was a self-destructive reaction in many directions, all of which collapsed.

On a more technical level, the RAF hoarded its best equipment in the United Kingdom, especially the Spitfires. Thus, it fought its campaigns with a limited effective force, notably Hurricanes; its bomber forces were untrained in ground support tactics, ineffectively armed for that purpose, and defensively weak. Even in the "colonial" campaigns in Greece and Malaya, the RAF could not compete. In part, that was because Whitehall never saw its opponents in those theaters as equals or better.

Suggestions for Further Research

The RAF is quite well covered in the official histories, the triservice volumes in the United Kingdom History of the Second World War, which includes subsets on grand strategy, operations, and the medical services and civilian infrastructure.

What has not been examined is the background, especially the technical side. For instance, what is the trail from cabinet decision to the ministries; to the manufacturers; to shipping; to unloading, preparation, and issue of aircraft and equipment; and then their sustenance, maintenance, wastage, and consumption in distant theaters? What about the recruitment, training, posting, and acclimatization of personnel?

The systems of communication have not been the subject of much study. Radar and Ultra have been explained to some extent, but there was a little noted technical watershed around 1942 in signals equipment.

Liaison with both the other services and allies was important for overall plans and operational necessities. These relations were heavily influenced by prewar concepts of the next war, by the mind-sets that ranged right to the top, and by perceptions of the enemy and estimates of his strength and intentions. All these need to be explored, as well as the nitty-gritty of daily routines.

Further research is also needed on the RAF in the period 1931–1942.

The logistics of maintenance, wastage, consumption, and supply—dull but basic subjects—are essential, as is the matter of gasoline and oil. More detailed studies of policy, planning, and distribution are also much needed.

Studies that would balance intelligence, assessments of enemy capabilities, and actual resources could be carried further. The readiness and serviceability rates of machines and squadrons could be analyzed from squadron and higher-echelon records found in RAF Forms 5401A and B in the Public Records Office (PRO; now the National Archives). The question of RAF concepts of war and allocation of motor transport could bear analysis, too. The evolution and training of debriefing officers and the assessment of the intelligence gathered also need study.

Finally, how did the RAF anticipate defeat, and how did it plan to manage impending doom?

Recommended Reading

The operational story was told briefly in the three popular volumes of *The Royal Air Force, 1939–1945,* by Hilary St. George Saunders and Denis Richards (London: HMSO 1953–1954, 2000). The interwar background has been covered by Robin Higham in a general context in *Armed Forces in Peacetime: Britain 1918–1940* (London: Foulis, 1963), and *The Military Intellectuals in Britain, 1918–1939* (New Brunswick, N.J.: Rutgers, 1966), and in a global context in *100 Years of Air Power and Aviation* (College Station: Texas A&M University Press, 2004). Further material on the operational side may be found in the official histories: T. K. Derry, *The Campaign in Norway* (London: HMSO, 1952); L. F. Ellis, *The War in France and Flanders* (London: HMSO, 1953); I. S. O. Playfair et al., *The Mediterranean and the Middle East,* vols. 1 and 2 (London: HMSO, 1954–1956); and S. Woodburn Kirby, *The War against Japan,* vol. 1, *The Loss of Singapore* (London: HMSO, 1957). Also important is F. H. Hinsley et al., *British Intelligence in the Second World War* (Cambridge: Cambridge University Press, 1978). Other official volumes can be located through the relevant chapters in

Higham's *Official Histories* (Manhattan, Kans.: Sunflower University Press, 1970) and *Official Military Historical Offices and Sources,* 2 vols. (Westport, Conn.: Greenwood, 2000). The original Air Ministry Air Historical Board (AHB) draft histories are now being reprinted by Cass and include not only *The Battle of Britain* (2000) but also the background volumes, *A Review of the Campaign in Norway* (PRO AIR 36/39) and *The Campaign in Greece* (PRO AIR 41/28). See also Air Ministry AP 3162, *Royal Air Force Glossary of Terms* (1950); Air Ministry AP 857, *Manual of Administration in the Royal Air Force* (March 1938), which does not cover the duties of staff officers; AIR 41/8, *Expansion of the RAF 1934–1939* (AHB c. 1943–1945); AIR 41/59, *Training;* AIR 41/39, *The RAF in the Bomber Offensive against Germany . . . 1927–1939* (AHB, n.d.); and AIR 41/71, *Manning: Plans and Policy.*

In addition to the civilian official histories, Sebastian Ritchie, *Industry and Air Power: The Expansion of British Aircraft Production, 1935–1941* (London: Cass, 1997); Colin Sinnott, *The Royal Air Force and Aircraft Design, 1923–1939: Air Staff Operational Requirements* (London: Cass, 2001); H. F. King, *The Armament of British Aircraft, 1909–1939* (London: Putnam, 1971); and Ian Lloyd, *Rolls-Royce* (London: Macmillan, 1978), cover technical aspects of the story, as do the hitherto confidential Air Ministry publications *Works* (London: HMSO, 1956), *Maintenance* (CD 1131, 1954), AIR 10/5570, *Signals V Fighter Control and Interception* (1952) (the problem in Norway, France, Greece, Crete, and the Far East was the want of radar sets and trained operators, sector controllers, and mechanics), and Higham's *The Bases of Air Strategy—Building Airfields for the Royal Air Force, 1915–1945* (Shrewsbury, U.K.: Airlife, 1998). Overseas airfields were the responsibility of the Royal Engineers; see their eight-volume history, R. P. Pakenham-Walsh, *The History of the Corps of Royal Engineers* (Chatham, U.K.: Institution of Royal Engineers, 1952).

See "Dickie" Dickinson, *Man Is Not Lost* (Shrewsbury, U.K.: Airlife, 1997), on air navigation in the RAF.

Also relevant are the three volumes of *The Royal Air Force Medical Services* by S. C. Rexford-Welch (London: HMSO 1945–1948), and

Air Ministry AP 3139, *Psychological Disorders in Flying Personnel of the Royal Air Force Investigated during the War 1939–1945* (London: HMSO 1947).

The intellectual, political, and economic background is covered in the first two volumes of the official *Grand Strategy,* volume 1 by N. H. Gibbs (London: HMSO, 1964) and volume 2 by J. R. M. Butler (London: HMSO, 1967). See also Barry D. Powers, *Strategy without Slide Rule* (London: Croom Helm, 1976).

On Dominion and Commonwealth participation, especially in the Middle and Far East, see Gavin Long, *Greece, Crete and Syria* (Canberra: Australian War Memorial, 1953), and W. G. McClymont, *To Greece* (Wellington, New Zealand: Internal Affairs, 1959); for Greece, see Higham's *Diary of a Disaster: British Aid to Greece 1940–1941* (Lexington: University Press of Kentucky, 1986) and Christopher Shores and Brian Cull, *Air War for Yugoslavia, Greece and Crete, 1940–1941* (London: Grub Street, 1987). Few of those who fought in these peripheral campaigns lived long enough to produce even wartime memoirs, but see E. C. R. Baker, *The Fighter Aces of the RAF* (London: Kimber, 1962), on Pattle, an unknown ace. On Norway there is François Kersaudy, *Norway 1940* (New York: St. Martin's Press, 1990), and S. W. Roskill's official *The War at Sea,* 4 vols. (London: HMSO, 1959–1961). On the air side of the battle over Flanders, see Robert Jackson, *Air War over France, 1939–40* (London: Ian Allen, 1974).

The ground-crew story has been told by Sir Philip Joubert de la Ferte in *The Forgotten Ones* (London: Hutchinson, 1961) and by Fred Adkin in *From the Ground Up* (Shrewsbury, U.K.: Airlife, 1983).

The interwar years are sparsely covered, although John Ferris in "Fighter Defence before Fighter Command: The Rise of Strategic Air Defence in Great Britain, 1917–1934," *Journal of Military History* 63 (1999), and Higham in "British Air Exercises of the 1930s," in *Proceedings of the 1998 National Aerospace Conference* (Dayton, Ohio: Wright State University, 1999), have pointed to the evolution of active air defense. John Ferris's "The Ultimate Enemy: The British Estimate of the Imperial Japanese Army, 1919–1941, and the Fall of Singapore," *Cana-*

dian Journal of History 28, no. 2 (1993): 223–56, examines intelligence and perceptions.

Useful adjuncts on the campaign in Burma are Henry Probert, *The Forgotten Air Force: The Royal Air Force in the War against Japan, 1941–1945* (London: Brassey's, 1998); Daniel Ford, *Flying Tigers: Claire Chennault and the American Volunteer Groups* (Washington, D.C.: Smithsonian, 1993); and Martha Bird, *Chennault: Giving Wings to the Tiger* (University: University of Alabama Press, 1987). See also F. Spencer Chapman, *The Jungle Is Neutral* (New York: Transworld, 1965).

Three articles on the Japanese air force are relevant, in addition to those mentioned in Mark Parillo's chapter 10: Rodrigo C. Mejia, "The Rise and Fall of the Imperial Japanese Air Force," *Australian Defence Force Journal* 141 (March–April 2000): 34–43; Alan P. Gentile, "Shaping the Past Battlefield for the Future: The United States Strategic Bombing Survey's Evaluation of the American Air War against Japan," *Journal of Military History* 64 (October 2000): 1085–112; and A. D. Howey, "Army Air Force and Navy Air Force: Japanese Aviation and the Opening Phase of the War in the Far East," *War in History* 6, no. 2 (1999): 174–204.

Conclusion

Stephen J. Harris and Robin Higham

One plane, one sortie, one bomb, and one high-value target destroyed; or, given technological advances, perhaps one plane, one sortie, *X* bombs, and *X* targets destroyed—with no friendly losses. Could any thoroughly modern major general (or air vice-marshal) ask for anything more?

Obviously, things are not that simple. Although stealth, global positioning systems, and increasingly sophisticated (and purpose-directed) hardware such as carbon filament bombs have made weaponeering more scientific, the precision is not absolute, and desired effects are not perfectly and predictably easy to achieve. What is the size or nature of the target? Will its destruction have any appreciable (or, better yet, decisive) impact on the course of the war, campaign, or battle? Adding the enemy's forces to the scenario, can one plane get there, bomb what it is supposed to, and return safely? In asymmetrical campaigns, do we really know what constitutes a high-value target in the first place?

These are not inconsequential questions, but the first sentence comes close to representing the philosophically ideal construct put forward by airpower theorists and advocates over the past ten decades. The air arm, they have argued, is uniquely capable of bringing victory more quickly and with fewer losses (at least to the successful attacker) than any other service—and it may, in fact, be able to achieve a war-winning, cataclysmic knockout blow independent of ground or sea operations.

Such was the philosophical essence of Giulio Douhet, Hugh Trenchard, and Billy Mitchell, the main theorists emerging from the First World War—an essence so powerful that nations that could not afford to build a bomber force (or had no targets for one) nevertheless became entranced by it, often at the expense of other elements of what we would now call air warfare. Their addiction to the offensive nature of airpower even in these early years is not difficult to comprehend. Flight has been part of the human dream for aeons, and its military application has likely been the dark side of that dream for almost as long. Leonardo da Vinci certainly knew what he feared when he sketched flying machines dealing death to cities, and Jesuit priest Francesco de Lama Terzi was so convinced that the ability to fly would cause cataclysm that he anticipated divine intervention to protect the human race from its own creative potential: "God would not suffer such an invention to take effect," he wrote in 1670, "by reason of the disturbance it would cause to civil government of men." Three hundred years later, the same vision was current, although it was then depicted as two scorpions in a bottle controlled by the reality of mutually assured destruction.

So far, the practitioners of airpower have only rarely achieved anything like the ideal of quick and lasting decisiveness. Apart from the two atomic bombs that produced Japan's capitulation, Israel in 1967 and the two Coalition wars against Iraq in 1991 and 2003 may be the best three examples, although there is sure to be debate about the last. Still, in each case, the balance of power was so one-sided that the conflict was practically no contest at all and thus perhaps not a legitimate test permitting credible conclusions. Otherwise, Hiroshima and Nagasaki excepted, defenses have been too strong, navigation has been too imprecise, bomb-carrying capacity has been slight and inaccurate, and explosive power has been too little; or, when the possibility became more likely, the political and social will to cause collateral damage amounting to thousands of casualties—mostly "innocent civilians" in major population centers—has been lacking.

Sir Arthur Harris spoke gleefully about "the destruction of German cities, the killing of German workers, and the disruption of civilised community life throughout Germany" in October 1943, and on Christmas Eve 1944, a reporter for the *London Sunday Express* explained how he felt upon finding "ten thousand [Germans] living like rats in cellars . . . it is good to think that what happened in Aachen goes on happening in every German town" (and "that on Christmas Eve," by an alleged Christian, fulminated J. F. C. Fuller). But Winston Churchill shrank from a series of Dresdens just a few months later, and Harry Truman, having approved the use of atomic bombs against Hiroshima and Nagasaki, found no compelling reason to do so again. Both recognized, as Clausewitz had observed, that when policy dissolved into hostility, destruction for its own sake became the objective, and both Churchill and Truman had sense enough of history and propriety (if not morality) to wish to avoid having the adjective "wanton" applied to their conduct of war. Something similar drove NATO as a whole to keep its gloves firmly on during the 1999 air campaign against Serbia. Although the senior air commander (and many of his air force subordinates) chafed at the restrictions placed on their conduct of operations, the consensus was that there was more to lose than to gain from unleashing the full potential of the allied air forces committed to the theater.

The theoretical ideal of independent air action that could win wars grew in the 1920s and 1930s partly because technology produced bigger and faster aircraft able to carry more ordnance farther, and partly as a reaction to the perceived dilution and diversion of airpower resources to do lesser things during the First World War: helping armies to win land battles and navies to control the seas. As that conflict showed, properly employed airpower was patently capable of doing both, and that reality, hammered home by sailors and soldiers not foolish enough to voluntarily forgo such an addition to their arsenals, left politicians with little choice. Although proportions varied by country and, as we have seen, within countries over time, airpower resources would be parceled out among a number of functions: "strategic" air attack, "frontal" attack (over land or sea), close support, reconnaissance, home ("strategic") air defense, frontal air defense, protection of strategic attack assets, interdiction (over land or sea), and logistics.

There were different force structures for different national circumstances, depending on individual design and production capabilities, threat assessments, and national goals. In some places, the air force was independent, or nearly so; in others, airpower was established within the army and the navy. Always there was competing doctrinal discourse, and whoever emerged as having the strongest voice mattered. Personality counted, too, at both the political and military levels. In Britain, Stanley Baldwin, Neville Chamberlain, and Sir Hugh Trenchard all mattered, but so did Sir Hugh Dowding and the minister for the coordination of defense. And, perhaps in a negative way, so did the lesser lights who for so long were responsible for army cooperation and antishipping and antisubmarine operations. (That these inferior officers led these organizations until partway through the Second World War was itself symptomatic of the thinking at the Air Ministry and on Downing Street.) Adolf Hitler, Ernst Udet, and Hermann Göring mattered in Germany; Ludomil Rayski and Jozef Zajac mattered in Poland.

No matter which doctrinal principles were in the ascendant, however, there was a kind of universal assumption that airpower had to be present everywhere. The axiom that followed was that once present, even in penny packets, airpower would make a critical difference to any military engagement. In short, the theoretical or philosophical ideal of strategic omnipotence percolated down to the smallest of land or sea campaigns. Somehow, as we have seen, a handful of B-17s stationed in the Philippines was assumed to be a strategic, operational, and tactical deterrent to the Japanese government and military. Somehow, despite an immature aircraft industry and limited budgets, the architect of the Polish air force from 1926 to 1938 thought that it could be strong in all its facets. Somehow, a few Gladiators and Hurricanes flying off frozen lake beds were to be important factors in Norway. And somehow, for an important few years in France, it was thought that one aircraft type could do everything.

The chapters in this book are about the falseness of the axiom—and the complexities of providing practical, concrete, real-world underpinning to doctrinal theory. Flaws, some of which were discernible at the time (and, it is argued, therefore should have been discerned), abound in these pages, but the reasons for these flaws—the essence of defeat and fall, the inability of air forces to operate offensively and defensively in-theater—are not as amorphous as one might think, despite the variety of cases examined. Robin Higham posits a plethora of possibilities in his introduction, and how they resonate likely depends on how individual readers' minds function. Still, some things seem perfectly clear.

To the aggressor go the spoils—at least in the early stages of a conflict, and if certain essentials are gotten right. This is an argument for the first strike by an air force better prepared for war because its government has dictated the pace—has worked its way inside the adversary's decision-making cycle at the political or grand strategic level. Germany, Japan, Israel (1967), Egypt (1973), and perhaps Argentina were all readier for the first day of war than their opponents, and all of them did reasonably well in the first hours, days, or months because they were better prepared. But as our authors have shown, early success does not guarantee ultimate success—first strikes can be risky. Indeed, of the examples noted above, only Israel (1967) had the wherewithal to maintain momentum and superiority, while Egypt, Germany, and Japan all wound down as their opponents became stronger.

Space undoubtedly played a role in these cases. Although Egypt had made territorial gains in the first days of the 1973 war, these were inconsequential and tradable for only a limited amount of time. Japan and Germany, by comparison, had overrun what would become a huge buffer zone that bought considerable time for each. The Luftwaffe was not decisively beaten in the west until February 1944, and even a year later, when little was being produced and less was being shipped from factories to the field, and the Allies had virtual air superiority over the Reich, the German army was still capable of fighting tenaciously. On certain days, even the air force was able to score well against both day and night bombing raids and against both the Soviet and Anglo-American armies. This legacy of capability did not amount to routine effectiveness, but neither did the lack of customary air support cause the Wehrmacht to put down its weapons the way the army (and society) of

November 1918 had given up the struggle. The Japanese, similarly, took more than four years—and submarine blockades and two atomic bombs—before giving in. Such were the patterns of Axis and Allied waves.

Argentina was likely a special case. Although its armed forces as a whole were better prepared and better positioned for the Falklands adventure than their British enemy, it was essentially left to the air force—the least well prepared of the three services for the particular operations at hand—to undertake the brunt of the fighting once the opponent was in-theater. It was another earned British victory to be sure, but one heavily influenced by original Argentinean sins: going to war before there were winning conditions (probably inevitable), misunderstanding the international reaction, and failing to be decisive in the defense of audacious aggression.

Nobody consciously plans to lose, and given the ever-present possibility (or hope) of decisiveness, to a greater or lesser extent, all the aggressors considered here anticipated quick victories. But things happen when seemingly irresistible force meets unexpected resistance or unforeseen circumstances. The Luftwaffe was clearly built to help the German army win continental land battles—a sensible priority for a continental land power—and it did so handily in Poland, the Low Countries, France, and the initial battles against the Soviet Union. Norway was a nearer-run affair because the Germans, too, dispatched a penny-packet contingent to the periphery. Still, with a better focus and a more stable and secure infrastructure (the peacetime bases of the Norwegian air force), the German air force seized the initiative and accomplished significantly more than its opponents did.

However, the prewar thinking, planning, design, and production that created such a powerful tactical and operational tool in 1939 had come at a price: strategic attack and long-range antishipping components were lacking from the Luftwaffe's repertoire. The hope, following the Battle of France, that it could achieve worthwhile results on the far side of the Channel—or, later, the far side of the Don—was perhaps better founded than the American hope that a few B-17s in the Philippines would deter (deflect or defeat) the Japanese, but in the end, the Luftwaffe, like the U.S. Army Air Force, was asked to do more than it realistically could. The Japanese, similarly, built their army and navy air arms to support these services and did well in their initial battles, but they too were confronted with the question of "what next?" after achieving initial victories in the Pacific. Although it may have been impossible to prevent the mass firebombing raids on Tokyo, it is significant that twice in August 1945 single B-29s flew over Hiroshima and Nagasaki totally unthreatened by Japanese air defenses. Whatever thinking the Japanese had done about "what next?" it had not gotten very far over forty-four months.

In the German and Japanese cases, and those of Egypt and Argentina as well, their failure to be decisive in the early going sowed the seeds of their eventual fall. The Luftwaffe had to try to become an air force capable of providing strategic air defense and producing strategic offensive power—while maintaining its operational competence in army support—all at once and while under increasing attack. Thus, it found itself in a situation that the French, Poles, Americans, Egyptians (1973), Argentineans, Soviets (1939–1941), and British (in the four campaigns studied in chapter 11) would have recognized. It had to play catch-up while in extremis. The French, Poles, Egyptians, and Argentineans never succeeded; the others did, largely because of space. American and Soviet production facilities were beyond the Luftwaffe's range; this was also true of the British aircrew training system and its production capability, once the Germans had begun to focus on their eastern enemy. Although the Germans may not have been aware of all the shadow factories in the United Kingdom, the locations of the main Vickers, Supermarine, Westland, Hawker, A. V. Roe, and Handley-Page facilities were known, and none of them was beyond German reach. But when the second "Little Blitz" was conducted, they were not the main point of the effort. Later, when Bomber Command and Eighth Air Force operations began to have effect, the fact that the Luftwaffe did so little to interfere through counterforce attacks was yet another failure, and a surprising one for an air force so conscious of the importance of knocking out its aerial opponents in 1939–1941.

Of the "catch-up" air forces, the RAF of 1939–1941 was clearly further along in the developmental cycle than the rest, even though it was not yet able to meet all its doctrinal and policy imperatives. (Again, the Argentine air force may have to be considered a special case, having been thrown into a contest for which it had made no preparations whatsoever.) British air defenses had been well organized since 1918, although there were still shortages in first-line aircraft (it required four-engine bombers and better navigation techniques to give its offensive doctrine a chance), and the Air Ministry had a reasonable idea of the more important targets in the German war economy. But just as the Germans had let strategic attack and antishipping considerations fall by the wayside, the RAF had neglected practically everything except strategic attack and home air defense, despite the political commitment to support land forces in France. Lacking numbers, equipment, stable infrastructure, and mobility; having failed to solve the problem of how best to support the army against a modern mechanized enemy; and then being let down by armies that themselves had not figured out the latter, the RAF in France had little chance of being effective. Neither did the bomber squadrons based in England.

Service doctrine that is not in harmony with government policy is likely to produce circumstances in which air forces will fail; government policy made in isolation of service capabilities tends to do the same. Avoiding such dissonance is not easy, even where think tanks abound. The United States in the 1960s always had superior airpower available over Southeast Asia during the Vietnam War, for example, and it was always capable of operating both offensively and defensively in the theater—the inability to do so being one of our definitions of "defeat" and "fall." But the deployed capability (and the aspirations of airmen) seemed out of synch with the guerrilla war in the south that marked Lyndon Johnson's presidency, and with the desire to find a political exit strategy (amounting to something less than victory) that marked Richard Nixon's. The great irony of Vietnam, in other words, was that when the gloves began to come off, albeit in a very controlled way, what airpower accomplished was to permit the United States to withdraw "with honor."

Perhaps the main lesson from the studies in this book is that the ends must be matched to the means in the short term, and when national survival is already at stake. To do otherwise is to risk frittering away resources on very long odds when there are more critical things to achieve. Conversely, the means must be matched to the ends in the long term, when there is time to think and plan. As all the chapters in this book indicate, failure to do the math beforehand has left air forces in the precarious position of having to fight the wrong battle at the wrong time, given their equipment, training, and resources. More often than not, however, the proper relationship between ends and means has been the counsel of perfection precisely because of the seductiveness of the ideal concept of airpower. Its offensive-decisive potentialities far surpassed its capabilities for most of the first century of flight—reach has routinely exceeded grasp—but it has been much too easy to hope and expect otherwise. "If airmen were like laboratory rats running a maze," Bernard Brodie observed in 1959, "they would seek to repeat success and to recoil from frustration. They would now be all in favor of tactical as against strategic uses of air power. But being instead very human, and knowing the power of nuclear weapons, they have remained intensely loyal to the original strategic idea"—that latent vision that Michael Sherry sees as "an abiding faith in the ritual of liquidation."

Politicians, being equally human, are no less susceptible to, no less seduced by, the point that the "original strategic idea" has been embraced in full even in countries with little if any indigenous design and manufacturing capability, no coherent system of public technical education, unhelpful geography, and major powers as likely adversaries. How else to explain Polish aspirations in the 1930s, which came at the expense of an effective air defense? At the same time, notions of weakness in comparison to others' perceived strength in the air have produced paralysis and fatalistic despondency. Neville Chamberlain was right to recognize and conclude that, given the weakness of Fighter

Command in 1938, German bombers would always get through; the RAF could not have won 1940's Battle of Britain in 1938. But the Wehrmacht of 1938 was not the Wehrmacht of 1940 either: it was not in possession of France and the Low Countries (and the airfields there), the Soviet Union was not yet a signatory to a nonaggression pact with the Reich, and the Czech armed forces were still in existence and potential allies.

Reading history backward, from the known, Chamberlain's policy, influenced by his fear of airpower, can at least be understood; it is far more difficult to comprehend the British and French belief that not bombing Germany after 3 September 1939 had any validity as the basis of national policy. Whether Anglo-French defeat in the west was inevitable, in either 1939 or 1940, is a matter of debate, but doing nothing to interfere with the buildup of Hitler's forces there left the sequence of future events entirely in his hands. The withholding of airpower, in short, meant that it had no impact—a kind of self-inflicted fall. (Admittedly, there is a gray zone of contention here. Arguably, withholding airpower is sometimes the right political decision. The question is whether doing so in 1939 was the same as limiting the bombing over Serbia in 1999.)

The dispatch of air contingents ill suited to the concrete task at hand also fits here. British expeditions in Norway, France, Greece, and Malaya are probably the clearest examples presented, although in the first three cases there was one small redeeming feature: the RAF would not be fighting the enemy's main body. Still, there were losses for little gain. Airmen involved in the Battle of France, meanwhile, could always point to the collapse of the armies around them as the overarching reason for their fall there, but the lack of radar, doctrine, and aircraft designed for the task rendered their efforts little more than a forlorn hope. It is difficult to accept that the air effort to win a campaign that the British had worried about for years—and that was serendipitously taking place on someone else's soil—was so poorly planned, supported, and executed. Very few countries have the opportunity to wage a decisive battle for national survival on foreign soil without having to fight for that buffer zone in the first place. Perhaps, then, the only redeeming feature of the RAF's conduct in France was that the Blenheims and Fairey Battles lost there were not a real blow to overall British air strength; the pilots and ground crews, however, were.

Tactics is for amateurs; logistics (and infrastructure) is for professionals. This epigram, or some version of it, is a well-worn tool used to persuade neophytes that military history involves more than the battlefield when the guns are firing. Because the concept of airpower is so easy—combat decisiveness is dramatic, after all, whereas supply and repair and facilities construction are not—the hard, practical importance of logistics has been easy to gloss over for both government and military planners alike. Indeed, the neglect has been endemic. Ignoring at least some of the complexities of design, testing, production, training (both air and ground crew), basing, and supply was common to all the outright losers (dead ducks) in this study—and, arguably, for all the phoenixes as well.

Underestimating the need, time, or industrial competence or capability required to keep pace with adversaries is a common component of defeat and fall, regardless of whether a country is self-sufficient or reliant on others for supply. (Whether the latter are able or willing to provide equipment that is at least comparable to an opponent's best is an additional complication for dependent air forces.) As used here, "need" is fundamentally different from "time" and "competence" or "capability," and it often reflects ostrich-like behavior at the political level and deficient thinking by the military. In either case, the consequences are clear: early defeat or, once under pressure, the temptation to maximize frontline strength by continuing serial production of older aircraft that are already outflown and outgunned. Interestingly, there appears to be a link between underestimating one's own infrastructure requirements and failing to recognize one's opponent's vulnerabilities in this area. The Germans (in the west), Japanese, and Argentineans all faced enemies who had to move much of their warfighting potential across thousands of miles of ocean, but in each case, besides being ill equipped to intervene in what should have been a predictable campaign, the air arms remained wedded to attacks on warships, rather than on supply ships.

Getting time and competence or capability projections wrong may reflect deficiencies in thinking and planning as well as engineering or technical surprises (foreseeable or not), but such errors may also be delusional, such as when industry is asked to design and produce more than it can realistically be expected to provide according to an impossible schedule. The impacts of faulty assessments differ. For example, that the Handley-Page Halifax was never quite as good as the Lancaster, or that neither could comfortably be fitted with .5-inch defensive armament, was one thing; they were produced in large numbers and flew on operations, and it is debatable how many more would have survived encounters with night fighters had they had better armament. That the Japanese were so late in thinking about self-sealing tanks was an oversight of a different magnitude; that the Germans placed faith in (and spent time and effort on) the so-called Amerika bomber when they were under increasing air attack was another miscalculation—one exacerbated by the folly of drafting skilled workers into the armed forces.

Although allowances can be made for the unforeseen in terms of design and production, it is more difficult to understand the neglect of bases and their affiliated infrastructures: repair and maintenance facilities, fueling, supply, and aerodrome defense. Compared with these, the sharp end of an air force is extremely small—and extremely fragile when they are lacking. Air forces, in short, cannot live off the land. Yet in several of the cases examined here, air contingents deployed away from their home bases were expected to do almost that. Sometimes, of course, the risk may be worth it. For example, as long as their enemy was the Italian air force, the RAF in Greece was likely to survive its meager surroundings and still be effective, despite limitations on sortie rates. (That says more, perhaps, about the Italians than it does about the RAF.) The RAF's situation in Malaya was obviously better than that in Norway—there was a purpose-built system of airfields in the former, although resupply was tenuous, given the limited road and rail networks. This was also true, before the fighting, of the situation in France, which was appropriate to a defensive campaign on a stable front near the German border. Polish provision of forward and "secret" airfields allowed its air force to survive the initial onslaught but simply prolonged the agony of defeat. The Soviets created their own vulnerability by adopting a strategy of forward defense; their fundamental recognition of the importance of infrastructure was the movement of industry and training beyond the Urals and the return of masses of skilled workers from the Red Army to the factory floor after the initial panic of June 1941. All told, the Japanese were the greatest culprits, having failed to grasp the importance of being able to build airfields quickly to give their airpower increased flexibility and mobility.

The sortie rates achieved by the Israelis in 1967 were startling—a force multiplier wholly based on infrastructure. But because it is so easily overvalued, such professional attention to detail has sometimes frustrated unknowledgeable citizenries and produced unhelpful public-relations problems. There might have been a little carping about a long buildup to the two Iraqi wars, but the extreme care taken to ensure that the Italian government would not object to operations during the 1999 air war over Serbia (and, in protest, close its bases to the NATO squadrons deployed there) was one reason why the campaign could not be waged as most airmen, and likely most Americans, would have preferred.

Mismanagement of human resources is another major common denominator of defeat and failure. It has qualitative and quantitative aspects, each impinging on the other, and is often rooted in the nature of the society. Internal societal structures in Egypt meant, for example, that the talent pool for both air and ground crews was limited—certainly a smaller proportion of the population than in Israel—and both the native and borrowed (Soviet military) culture tended to stifle initiative, meaning that the best use of available aircraft was not necessarily assured. Germany and Japan, in contrast, ensured that they were ready at the outset—their air forces were highly trained and motivated—but having failed to think through the implications of not achieving a decisive success quickly, they overlooked the human investment necessary to wage a long war in the air. Both squandered their most experienced pilots and then had to rush increasingly inexperienced replacements into the fight, with predictable results. Starting late, and still placing stock in strategic attack, the RAF had too few fighter pilots undergoing training to survive a prolonged Battle of Britain, but it had already taken steps to rectify the situation in the long term. The Empire Air Training Scheme agreed to in December 1939 was a well-considered effort to facilitate training throughout the Empire and the Commonwealth and to ensure that output went where it was needed and in the right proportion. British air operations from 1941 on were possible only because there had been clear, unambiguous, conscious recognition of the need for numbers, and because the pressure to reduce the quality of training was resolutely resisted. In Canada, at least, it was the army that complained that too many of the best and brightest were joining the Royal Canadian Air Force.

The importance of the ability to sustain operations should be self-evident, but the problem still seems to exist, albeit in a slightly different form. Since the end of the Cold War, the major air forces of the West have all seen a significant decline in their regular, full-time, active-duty strength, yet they have been employed on active operations to a greater extent than anyone could have predicted when there were thoughts of the peace dividend. Not only has that put pressure on regular units—perstempo as part of optemo—and raised questions about the sustainability of operations; it has also led to a far greater reliance on reservists to flesh out "war establishments." Whether justified or not, friendly-fire incidents in campaigns where minimal collateral damage is a measure of excellence have raised questions about the readiness of the latter.

Although aircrews (and especially pilots) usually take pride of place, air forces are materiel organizations utterly dependent on complex understructures. When these are degraded, airpower cannot be effectively projected. The essays in this collection have examined the crushing, quick, and total defeat of air forces (Iraq, Poland, France, Egypt 1967), defeat in quick campaigns and "sideshows" (Malaya, Philippines, Russia, Argentina, Egypt 1973), and long-term disintegration. German and Japanese victories in France, Russia, and the Far East could not be held because they failed to destroy enough of the enemy's main body to prevent its regrowth. Dramatic Israeli military victories have not produced long-term stability or security. Near equals have been featured in some; unambiguous disparities in others. "Generalship," or its air force synonym, is generally absent from the discussion, although there is a lot of "planning" and doctrine, which may in fact be the heart of air force generalship (Dowding's firmness in not committing additional forces to France is likely the best example). Rarely have aircrews simply laid down their arms and withdrawn from the fight, no matter what the odds or loss rate. "Not ready," or some other construction with the same meaning, seems to be the conclusion that can be used for every fallen air force—not ready because of quantity or quality of aircraft, aircrews, or both; or because of deficient doctrine for the campaign at hand; or because the air force had (or chose) to operate without taking care of materiel infrastructure. These suggest that the fall of an air force is the result of long-term failings, not an immediate failure "on the day" by an air arm that is essentially ready for its allotted role.

Contributors

Anthony Christopher Cain is an active-duty colonel in the U.S. Air Force. He has flown more than 3,000 hours as a B-52 navigator and bombardier. A veteran of Operations Desert Shield and Desert Storm, he earned the Distinguished Flying Cross and two Air Medals while flying twenty-six combat missions against Iraq in 1991. He earned his PhD in military history from the Ohio State University in 2000. Presently he serves as the dean of education at the U.S. Air Force's Air Command and Staff College. He lives in Montgomery, Alabama, with his wife, Mechieal, and their three children, Jessica, Micah, and Ryan.

James S. Corum received a BA in German and history from Gonzaga University (1975); enrolled at the University of Heidelberg (1974–1975) and passed the German University Sprachzertifikatprüfung; and received an MA in history from Brown University (1976), a Litt M in history from Oxford University (1984), and a PhD in history from Queen's University, Canada (1990). A specialist in airpower and military history, he was a professor of comparative military studies at the U.S. Air Force School of Advanced Airpower Studies, Maxwell Air Force Base, Alabama, from 1991 to 2005. He is now at the U.S. Army Command and General Staff College, Fort Leavenworth, Kansas. Corum's publications include *The Roots of Blitzkrieg: Hans von Seeckt and German Military Reform* (University Press of Kansas, 1992)—winner of the New York Military Affairs Symposium Best Book Award of 1992; *The Luftwaffe: Creating the Operational Air War 1918–1940* (University Press of Kansas, 1997)—a main selection of the History Book Club; and *The Luftwaffe's Way of War* (Nautical and Aviation Press, 1998).

René De La Pedraja received his PhD in history from the University of Chicago in 1977. He resided for many years in Latin America but is currently a professor in the Department of History at Canisius College in Buffalo, New York. De La Pedraja has published numerous books and articles on Latin America, most notably in the area of business history. His early interest in merchant shipping eventually brought him to the field of military history. During the last six years, he has been researching and writing a book on the wars in Latin America from 1900 to 1941, which is expected to be published in 2005. De La Pedraja is already at work on another project—a book-length history of the Peruvian army.

Stephen J. Harris is the chief historian, Directorate of History and Heritage, National Defence Headquarters, Canada. He specializes in air history and was one of the coauthors of *The Crucible of War: The Official History of the Royal Canadian Air Force III* (1994) and of a chapter on the Royal Air Force in *British Military History: A Supplement to Robin Higham's Guide to the Sources* (1988). As well as managing the Canadian official histories program, he is currently working on a history of the 1999 air war over Kosovo.

Robin Higham was born in London in 1925 and served as a pilot in the Royal Air Force Volunteer Reserve (1943–1947). He graduated cum laude from Harvard College (1950) and received an MA from Claremont Graduate School (1953) and a PhD from Harvard University (1957). He became an American citizen in 1954. A leading aviation history author as well as the editor of a number of contributed works, his latest work is *100 Years of Air Power and Aviation, 1903–2003* (Texas A&M University Press, 2003).

David R. Jones has long been involved in Russian and Soviet studies and has been the director of the Russian Research Center of Nova Scotia, Halifax, since 1977. He received a PhD from Dalhousie University. Jones has served as editor of several encyclopedias in the field and was on the editorial board of the *Journal of Soviet* (now *Slavic*) *Military Studies.* He contributed to *The Military History of Tsarist Russia* (2002) and *Russian Aviation and Air Power* (1998).

John H. Morrow Jr. is Franklin Professor of History at the University of Georgia. He specializes in the history of modern Europe and of warfare and society. He is the author of *The Great War in the Air: Military Aviation from 1909 to 1921* (1993), *German Air Power in World War I* (1982), and *Building German Air Power, 1909–1914* (1976). He also wrote the chapter on airpower in the *Oxford Illustrated History of the First World War* and edited *A Yankee Ace in the RAF: The World War I Letters of Captain Bogart Rogers* (1996). His most recent work is *The Great War: An Imperial History* (2004), a comprehensive history of the First World War from its origins through its aftermath.

Mark Parillo of Kansas State University teaches aviation history and courses on World War II and general military history. He has written *The Japanese Merchant Marine in World War II* (1993) and several articles and book chapters on the Pacific war. He has been the newsletter editor for the World War II Studies Association since 1993. He is currently writing a sourcebook on Pearl Harbor for undergraduates and a monograph on land transportation in Southeast Asia during World War II, including a reexamination of the Burma Road.

Michael Alfred Peszke is a psychiatrist of Polish descent who has long written about Polish military and air force history. He is the author of a number of important articles in the field, including "Historiographic Problems in the Study of the Polish Air Force in World War Two" (*Journal of American Aviation Historical Society* 18 [1973]: 58–64); "The Operational Doctrine of the Polish Air Force in World War Two: A

Thirty Year Perspective" (*Aerospace Historian* 23 [1976]: 140–47); "Prewar Polish Air Force: Budget, Personnel Policies and Doctrine" (*Aerospace Historian* 27 [1981]: 168–89); and "The Forgotten Campaign: Poland's Military Aviation in September, 1939" (*Polish Review* 39 [1994]: 51–72).

Brian R. Sullivan graduated from Columbia College and was commissioned in the Marine Corps in 1967. He served in Vietnam, receiving a Silver Star and a Purple Heart. He received his PhD in modern European history from Columbia University. Sullivan then taught modern Italian history and twentieth-century European military history at Yale University (1984–1988), was Secretary of the Navy Fellow and member of the Strategy Department at the Naval War College (1988–1991), and was a senior research professor at National Defense University (1991–1997). During the Gulf War, he advised the Defense Department on special operations, deception, and psychological operations. Since 1998, Sullivan has worked as an independent consultant on national security and defense-related issues, while continuing to research and write about Italian military and naval history. He coauthored *Il Duce's Other Woman,* a biography of Margherita Sarfatti, Mussolini's longtime adviser. It was awarded the American Historical Association's Marraro Prize for the best work on Italian history in 1993.

Osamu Tagaya is the son of a former officer of the Air Technical Arsenal of the Imperial Japanese Navy. He has devoted a lifetime to the passionate pursuit of his interest in the history of military and naval aviation, focusing particularly on the air forces of imperial Japan. Using his fluency in both Japanese and English, he has endeavored to bring to Western readers a better understanding of Japanese aviation through his books and articles on the subject, including works for the Smithsonian Institution and a number of commercial publishers and magazines. He is an investment banker and financial investor by profession.

Index